社区建筑
In the Community

汉英对照
(韩语版第365期)

韩国C3出版公社 | 编
党振发 楚立峰 李思逸 蒋丽 张琳娜 周一 | 译

大连理工大学出版社

4 社区建筑
公共建筑与机构

006 社区规模建筑的大变化 _ Douglas Murphy

012 本迪戈图书馆 _ MGS Architects

022 斯泰普尔顿分馆 _ Andrew Berman Architect

032 Vennesla图书馆和文化中心 _ Arkitektfirma Helen & Hard AS

046 Cooroy图书馆 _ Brewster Hjorth Architects

054 巴达霍斯精品艺术博物馆 _ Estudio Arquitectura Hago

066 沃特福德中世纪博物馆 _ Waterford City Council Architects

076 手工造纸博物馆 _ TAO

090 澳大利亚恐龙时代博物馆 _ Cox Rayner Architects

100 个性化建筑

100 机构模式 _ Alison Killing

106 Råå日托中心 _ Dorte Mandrup Arkitekter

114 农场幼儿园 _ Vo Trong Nghia Architects

126 Pies Descalzos学校 _ Mazzanti Arquitectos

136 Zugliano的某所学校 _ 5+1AA Alfonso Femia Gianluca Peluffo

144 El Guadual儿童早期发展中心 _ Daniel Feldman & Iván Dario Quiñones

158 木质牙科诊所 _ Kohki Hiranuma Architect & Associates

168 哥本哈根癌症防治与康复中心 _ Nord Architects

178 利迈健康中心 _ Atelier Zündel Cristea

188 建筑师索引

4 In the Community

The Public and Institutional Buildings

006 *The Changing Face of Community-Scale Architecture _ Douglas Murphy*

012 Bendigo Library _ MGS Architects

022 Stapleton Branch Library _ Andrew Berman Architect

032 Vennesla Library and Culture House _ Arkitektfirma Helen & Hard AS

046 Cooroy Library _ Brewster Hjorth Architects

054 Badajoz Fine Arts Museum _ Estudio Arquitectura Hago

066 Waterford Medieval Museum _ Waterford City Council Architects

076 Museum of Handcraft Paper _ TAO

090 Australian Age of Dinosaurs Museum _ Cox Rayner Architects

100 The Individuality and the Institution

100 *Institutional Patterns _ Alison Killing*

106 Råå Day Care Center _ Dorte Mandrup Arkitekter

114 Farming Kindergarten _ Vo Trong Nghia Architects

126 Pies Descalzos School _ Mazzanti Arquitectos

136 School Complex in Zugliano _ 5+1AA Alfonso Femia Gianluca Peluffo

144 El Guadual Early Childhood Development Center _ Daniel Feldman & Iván Dario Quiñones

158 Timber Dentistry _ Kohki Hiranuma Architect & Associates

168 Copenhagen Center for Cancer and Health _ Nord Architects

178 Medical Care Center in Limay _ Atelier Zündel Cristea

188 Index

社区建筑

In the Co

本期要为读者介绍的是社区公共建筑。这里包含了四种类型的公共建筑：博物馆、图书馆、医疗建筑和教学建筑。C3丛书之前曾经探讨过类似的建筑，社区本身作为一项功能融入其中，成为其存在的理由，例如"都市与社区"（41期）主题中介绍的社区中心。然而，本期将要探讨的公共建筑则具备了更加细化的功能。它们的规模相对较小，为所在地区或居民服务。

本期主要涉及为地方社区建造的两种小规模公共建筑。

第一章"公共建筑与机构"中介绍博物馆和图书馆。借由网络途径就能够实现的日益增多的知识资源共享，以及日渐涌现的大规模博物馆建筑正在严重威胁小型图书馆或者博物馆的基本生存问题。这一章将向读者展示小型社区建筑是如何回应这些挑战的。它们也许不能像大规模建筑那样成为地标，但是它们能够更深层次地融入居民的生活之中，与社区发展友好的关系。

This issue introduces public buildings for the community. There are four types of public buildings here: museums, libraries, and health and education buildings. C3 has previously looked at cases in which the community itself became functionally subsumed into the building's raison d'etre, like a community center in the theme of "Community and the City" (#41). This time, however, public buildings that have more accurate functions are examined. They are relatively small in scale, being for local areas or residents.

This issue deals with two categories of small-scale public buildings for such local communities.

The first chapter "The Public and Institutional Buildings" introduces museums and libraries. The expansion of information sharing via the Internet and the emergence of more and more large-scale museums are threatening the very survival of small libraries or museums. This part shows how small community buildings are responding to such challenges. That is, rather than presenting themselves

公共建筑与机构 The Public and Institutional Buildings
个性化建筑 The Individuality and the Institution

mmunity

第二章"个性化建筑"谈及了与医疗和教学功能相关的建筑。

这两类建筑有一些非常重要的共同点。它们作为基础设施运营机构,都为当地社区提供必要的服务。与同类设施一样,它们也要为机构安排的组织活动提供空间,因此会显得缺乏个性化,风格千篇一律。

在这样组织有序的机构体系中,空间的安排反映了其注重效率的中心需求。然而,隐于建筑之中的个性元素也正因此而被发现。这一章节将向读者介绍几例通常乏于展示个性机会的建筑,是如何逐渐实现自我转变,与居民形成新关系的。

as landmarks as might large-scale buildings, they get involved in residents' life at a deeper level to develop friendly relationships with the community.

The second chapter "The Individual and the Institution" ranges over health and education buildings. These two types of buildings have a few, quite significant things in common. They function as facility operating institutions to provide essential services to the local community. Being such they also share any such facility's institutionally arranged organization of space, with its implications for lack of personalisation and monotonous style.

The arrangement of space in such rationally organized institutional systems reflects the centrality of the need to focus on efficiency. And yet, withal, the element of the personal is somewhere to be found within. This chapter introduces some cases in which buildings that used to lack opportunities for personalisation gradually transformed themselves, forming new relationships with residents.

公共建筑与机构
The Public and Instituti

当今世界,小型社区建筑的作用是什么? 20世纪,交通运输的发展使诸如购物商场一类建筑的功能大大集中。但近年来,社会变成了数字化社会,这一转变使另一类活动变得集中,即借助于互联网零售和社会传媒活动来改变人们与其居住的城市之间的互动方式。尽管人们对住房、工作空间和娱乐的需求一直很大,但可供选择的服务和功能越来越少。在本篇文章中,我们将研究一些小型或中型新建筑,每一座建筑都展现了解决建筑问题所运用的不同方式。

或许在功能方面,受到威胁最大的是小型公共图书馆。由于印刷资料数字化且从网上很容易获取的特征,图书馆的传统功能变得越来越不重要。尽管图书馆这一概念仍受很多大众喜爱,但作为新类型,图书馆必须以新形象示人,即给人们提供更多的空间,把休闲和保健设施融入到结构中。在本文中我们将了解不同的图书馆建筑,有的作为传统的延续,有的体现地方图书馆功能的新理念。

小型博物馆不易受欢迎的部分原因是因为一些著名的大型博物馆和收藏市场的大幅度扩张。如果说全球知名博物馆建筑是最后一代最引人注目的、最驰名的建筑,那么这种情况对小型的、使人感到更亲近的博物馆造成了什么样的影响呢? 我们来看一看某些新建的小型博物馆:它们与社区的关系比与那些更有魅力的"表亲"建筑的关系更复杂、更微妙,从而与图书馆诠释给人们的形象形成有趣的对比。

What role do small community buildings play in today's world? In the 20th century developments in transportation led to a great centralisation of functions in typologies such as the shopping mall, but in recent years the transition to a digital society is beginning to enact another centralisation of activity, with Internet retail and social media all promising to change the way that people interact with their city. While housing, workspace and entertainment all continue to have great demands, all manner of alternative services and functions are in decline. Here we will examine a number of different new small-medium sized buildings, each one of which points to different ways in which architecture addresses this problem.

Perhaps the most threatened function is the small public library. As printed information becomes digitised and more easily available online, the functions traditionally offered by these buildings become less and less vital. While there is great public affection for the concept of the library, as typologies they are having to reinvent themselves as more social spaces, often bringing in leisure and care facilities into their structures. We will look at a series of varied library buildings which range from continuation of the tradition to new ideas of how a local library should function.

Small museums are also vulnerable to a decline in popularity, partly due to an expansion of the market for blockbuster museums and collections. If large, globally branded museum buildings have provided some of the most dramatic and celebrated architecture of the last generation, how does this effect the design of smaller, more intimate museum buildings? Here we will consider a number of new small museum buildings which have more complex and subtle relationships to their community than their more glamourous cousins, and which offer an interesting contrast to the attitudes put forward by the libraries themselves.

本迪戈图书馆_Bendigo Library/MGS Architects
斯泰普尔顿分馆_Stapleton Branch Library/Andrew Berman Architect
Vennesla图书馆和文化中心_Vennesla Library and Culture House/
Arkitektfirma Helen & Hard AS
Cooroy图书馆_Cooroy Library/Brewster Hjorth Architects
巴达霍斯精品艺术博物馆_Badajoz Fine Arts Museum/
Estudio Arquitectura Hago
沃特福德中世纪博物馆_Waterford Medieval Museum/
Waterford City Council Architects
手工造纸博物馆_Museum of Handcraft Paper/TAO
澳大利亚恐龙时代博物馆_Australian Age of Dinosaurs Museum/
Cox Rayner Architects
社区规模建筑的大变化_The Changing Face of Community-Scale Architecture/
Douglas Murphy

社区规模建筑的大变化

在当今由数字化连接而成的世界，人们有时感觉到建筑的种类呈下降趋势。例如，在20世纪，主要街道因为室内商场的开发而遭到破坏，且街道旁的空置空间被商店挤占，车辆可以进入这些空间，由此达到了当地的一些小企业无法与其匹敌的效率。但是现在，21世纪技术的进步对这些实体店构成了威胁，迫使它们搬迁，只留下储存货物的仓库或外销产品的销售网络。有时人们会感觉到，他们唯一能看见的建筑就是自己的家或工作地，因为越来越多的活动都是在网上进行的。

或许最明显的衰退就是公共图书馆。自从有了文字作品，图书馆便出现了。公共图书馆起源于19世纪中叶，出现在英国和美国。它们当时是慈善和教育力量的标志。安德鲁·卡内基和帕斯莫·爱德华兹等实业家捐赠巨资给数以千计的图书馆，旨在"改善"大众的生活。他们认为这是社会富有阶层人士的公民和道德义务。后来，在20世纪的战后时代，图书馆成为与国家慈善相关的一类建筑。慈善家贯彻慈爱的理念，并且建造了许多大大小小的图书馆。

随着千禧年的到来，公共图书馆笼罩在过时的阴影中。大量档案资料变成了数字形式，严重削弱了图书馆的使用价值。即使尚未被迫关门，但已进入困境，在社会和技术环境千变万化的形势下，图书馆必须重塑其公共建筑的形象。在伦敦，David Adjaye于2005年提出了著名的"概念店"。它是一种通过引进各种教育和社会活动来重振图书馆的意义的重大尝试。显然，作为旨在扩大公共范围的机构，"图书馆"这个名字已经从机构的名单中去除。而且，或许也疏远了本身与人们在不同历史时期对公共信息的需求的关系。

当然，世界各地仍在有条不紊地建图书馆。但是，作为一种建筑，我们可以看出，它们正在努力地利用不同的方式来与当今的生活方式密切相关。由安德鲁·伯曼建筑师事务所在美国的斯坦顿岛建造的斯泰普尔

The Changing Face of Community-Scale Architecture

In today's digitally connected world, it can sometimes feel as though there is a constant decline in the variety of different types of buildings. For example, high streets suffered in the 20th century, due to the development of the indoor mall, whose blank serviced space into which any shop could insert itself, all accessible by car, led to efficiencies that small local enterprises couldn't compete with. But now, the technological advances of the 21st century are even threatening to remove even the physical environments of shopping, leaving nothing but warehouses for storage and the distribution networks which bring the products to out doors. It can sometimes feel as though the only buildings people encounter now are their homes and their work places, as more and more activities take place online.

Perhaps the most obviously declining typology is that of the public library. While libraries have existed for as long as there has been writing, in the mid-late 19th century, originating in the UK and the USA, the public library became a symbol of philanthropy and the power of education. Industrialists such as Andrew Carnegie or Passmore Edwards spent sizable amounts of their fortunes endowing thousands of public libraries for the "improvement" of ordinary people, in the belief that this was their civic and moral duty as wealthy members of society. Later, during the post-war era of the 20th century, the library was one of the architectural typologies most associated with the beneficent powers of the state, who carried on the philanthropic ideal, building many important buildings, both large and small.

But since around the millenium a cloud of obsolescence has been hovering over the public library, whose usefulness has been undercut primarily by ongoing transfer of much archive information into digital forms. If they haven't already been forced to close by now, they have been forced into the difficult situation of trying to reinvent their purpose as civic buildings in a changing social and technological landscape. David Adjaye's celebrated "Idea Store" of 2005 in London was a significant attempt to reinvigorate the typology by introducing all kinds of educational and social activities. Tellingly, the name "Library" was omitted from the institution, as an attempt to broaden its civic scope, but also perhaps to distance itself from the associations with a different historical need of the people to public information.

Of course, libraries are still regularly built across the world, but in the building as it exists now we can see many different ways of

顿分馆就是一个典型的例子。该分馆合并了1907年建的卡内基图书馆，对其内部进行了翻新，使扩建后的图书馆的规模是原来的两倍之多，且新老建筑相连。

建筑师们表示，"该图书馆被设计成开放、方便利用、使人感到亲近的建筑。"首先，它反映了一直未解决的方便利用的问题。出于种种原因，图书馆未能成为全体大众都可以利用的设施。在过去的十年中，图书馆所做的大部分工作就是增强其吸引力。无论图书馆规模的大小，通过木质的结构和某些裸露的饰品，它都给人一种亲近感。这种新式建筑回应了原有建筑的装饰风格，给人一种有机的、温馨的感觉。这种在街道一侧全部镶嵌玻璃、在某种程度较为正式的设计暗示出这类项目内在的"公共特征"。

最近，另一座图书馆竭力将现代设计手法融入到传统的建筑中。它位于挪威的文内斯拉小镇。建筑师海伦和哈德沿着这个小镇的主要大街建造了一座非常醒目的新图书馆。它面朝广场，以展现其作为传统的公共建筑的重要性。大型玻璃立面使建筑成为公共空间的延伸结构。面向街道的立面上的大面积阶梯式结构营造出一种增高的效果。图书馆内部设计更加新颖，穿过图书馆的、两侧的木质结构入口之间的空间布满了书架、照明设备、通风设备和其他服务设施。一个入口是常规的结构，而另一个入口形成了上空空间。这两个入口为一系列的曲形带状结构，带有圆形的墙角、奶油色的漫射光，给人以愉悦之感，具有复古未来的吸引力。人们途径此处都能感受到每一个构件独一无二的特色。木质结构彰显了斯堪的纳维亚的建筑传统。海伦和哈德的作品在风格方面又独具匠心，和地方环境紧紧融为一体。

在这些小社区的背景下，这种新图书馆的大胆建筑风格是常见的手法。它们不求庞大。建造为社区内的所有人带来益处的机构，这种尝试引领了其他类似的建筑手法。由MGS建筑师事务所在澳大利亚的维多利亚建造的本迪戈图书馆合并了之前的20世纪80年代的图书馆。这座老建筑被改造成一座低矮的二层建筑，带有大进深平面，由屋顶天窗进行照

trying to keep the typology relevant to today's lifestyles. A typical example of this would be the Stapleton Library extension by Andrew Berman Architects, on Staten Island in the USA. This project works with an existing Carnegie Library of 1907, with an interior refurbishment connecting to an extension over double the size of the original building.

The architects state that "the library is designed to be an open, accessible, and intimate building." Firstly, this reflects the ongoing problem of access for whatever reasons, libraries had come to be facilities whose users were unreflective of the population as a whole, and a large part of much library work over the last ten years has been the attempt to widen the appeal of the institutions. Despite the size of the building, a sense of intimacy has been achieved through the use of timber for the structure and certain exposed finishes, the new construction echoing the existing building's decorative work, and also having an organic warm quality. The scheme is fully glazed to the street, and is organised in a somewhat formal manner, hinting at the civic qualities inherent to a project of this kind.

Another recent library which attempts to add a modern twist to the traditional type can be found in the small town of Vennesla in Norway. Here, architects Helen and Hard have built a striking new library building along the town's main street. The library faces the square, asserting the traditional civic importance of the building, while its large glazed facade is there to allow the building to be seen as an extension of this public space, an effect which is heightened by the provision of extensive benching on the facade, facing the square. Inside, though, the building is more innovative. The bookshelves, lighting, ventilation and other services are all contained in the spaces created by gaps between two timber portals which stretch right across the building. One portal is structural, whereas the other creates the void spaces, but together they read as a series of curved banded structures with an agreeable retro-futuristic appeal, with rounded corners and milky-diffused light, each one unique as they pass across the irregular site. With its timber construction it can claim to be within the Scandinavian building tradition, but Helen & Hard's work is stylistically innovative, a strong addition to the local environment.

Within the context of small communities, the architectural boldness of these new libraries is a common strategy. Without relying on monumentality, the attempt to create an institution capable of providing something for all members of the community results in similar architectural approaches. The Bendigo Library by MGS Architects in Victoria, Australia, incorporates a previous 1980s library

明。这种设计是典型的超越了传统的图书馆的案例，其独特之处是它含有咖啡厅、艺术馆、志愿者服务台、研究设施和网络设施。

　　建筑师的目的是建造一座"灵活的、包罗万象的图书馆"。它也是一个会议交流场所，其设计选择松散自由的平面，但材料色彩明亮，设计语言生硬，这种将多种材料和形式融合到一起且不太在意结果的设计就是我们所说的"墨尔本学校"人员（诸如建筑师Lyons，Hassell或NH）的风格。他们的作品给人一种幻觉，但这种手法意在营造使人放松的空间。

　　在澳大利亚的其他地方，如昆士兰州的Cooroy，Brewster Hjorth建筑师事务所运用了类似的设计手法将一座新图书馆与地方环境紧密结合。新图书馆弯曲的两翼结构使小镇的主要大街与艺术馆产生层次变化（艺术馆场地原为工业用地），且其再次被视为社交中心。社区的所有人都聚集于此。咖啡厅、数码广场、休息厅、社交空间都采用五彩缤纷的、柔和的材料装饰，而图书馆试图吸引社区老少来到这处公共空间。

　　这些图书馆建筑的一个普遍特点就是不把这一机构变成永久性的纪念物。在这处小型社区团体的背景下，图书馆试图摒弃其过时的功能——在固定期限内借阅图书，而是成为类似于市政厅或社区中心的建筑，不断地将社交和娱乐功能融入到设计方案中。但是，如果我们来看另一种类型——与纪念物有很强历史联系的博物馆，我们就会发现在不同的小社区内建造的建筑是如何解决相似的问题的。

　　像图书馆一样，博物馆在19世纪的帝国时代得到了巨大发展。那时，改革者们试图给刚刚自我觉醒的大众提供教育和信息。与图书馆不同的是，现在，博物馆仍然那么强大，那么受欢迎。在很多情况下，它是建筑师的使命。通常，当博物馆处于主导地位，在全球都有重大影响，成为建筑与其藏品都同样重要的一个大型旅游景点时，其受欢迎程度便会达到顶峰。那么这种设计方式能够按照比例缩减吗？

　　在澳大利亚，澳大利亚恐龙时代博物馆是由Cox Rayner建筑师事务所设计的，它展现了一个完全不同的设计方法。这座建筑设有古生物展览区，是受一个出让土地来探索化石遗迹的农场家庭的委托。这座小建

斯泰普尔顿分馆，纽约，美国
Stapleton Branch Library in NewYork, USA

building, which has been reworked into a shallow two-storey building with a very deep plan lit by rooflights. The programme is typically expanded beyond a conventional library, and now features cafes, galleries, volunteering services, research and internet facilities.

The architects' stated aim is for a "flexible and inclusive library", a space for meeting and exchange, and its design sets out to achieve this not only with its loose and freely formed plan, but with its brightly coloured material choices and angular design language. The carefree juxtaposition of so many materials and forms is typical of what we might call the "Melbourne School" (architects such as Lyons, Hassell, or NH), whose public work sometimes borders on the psychedelic, but here the approach works towards creating a space which is relaxed and informal.

Elsewhere in Australia, in Cooroy, Queensland, Brewster Hjorth Architects use a similar design strategy to tie a new library into its local environment. Consisting of two curved wings which bridge a change in level between the main street of the town and an art gallery (which occupies a site previously occupied by industrial uses), the new library is again conceived as a social hub, where all parts of the community come together. Again, cafes and digital suites, lounges and social spaces are dressed in colourful, friendly materials, as the library attempts to attract the youth and elderly of the town into its civic spaces.

A common strand in these library buildings is an attempt to de-monumentalise the institution. Within the context of smaller community groups, they attempt to remove what might be seen as an archaic function of libraries – the borrowing of books for a finite period of time – and to become something more akin to a town hall or a community centre, frequently incorporating social and entertainment functions into their programmes. But if we consider another typology – the museum – which also has a strong historical relationship to monumentality, we can see how different small community oriented buildings approach similar problems.

Like libraries, museums are a typology which experienced a massive growth in the imperial age of the 19th century, when reformers attempted to provide education and information to the newly self-conscious masses. Unlike library, however, the museum today is still a powerful and popular experience, and in many cases it is the ultimate commission for an architect. But often the museum is at its most popular when it is a dominant, globally significant building, a large scale tourist attraction whose architecture is as important as its contents. Can this approach be scaled down?

澳大利亚温顿恐龙时代博物馆，昆士兰
Australian Age of Dinosaurs Museum in Winton, Queensland

腾冲手工造纸博物馆，云南，中国
Museum of Handcraft Paper in Tengchong, Yunnan, China

筑建在峡谷之上，几乎和从岩石中生长出来的一样。呈现出一种它是由（周围挖掘的土制成的）夯土墙建成的强烈效果。与使人感到亲近、五彩缤纷，且与周围环境分开以成为当地焦点的图书馆不同，这座博物馆几乎不像是一处场所。

恐龙博物馆完全由那些无偿奉献时间的顾问和承包商们建造而成。他们聚到一起，共同致力于为他们认为值得的项目服务。使用类似的合作设计方式的建筑还有中国云南省的手工造纸博物馆。它由TAO·迹建筑事务所设计。该博物馆只用了当地的材料和技术，包括在展会上展出的剪纸技术，来试图与环境分离开来。像许多近年走出中国的更多的建成建筑那样，这个项目完全采用了当代设计手法（不规则的几组盒式房间、无框架的玻璃，以及朴素的材料），并且让它们直接与本地特点相结合，创造出一座带有本地原生态特色的建筑。在这里，建筑师必须创造一套复杂的实体模型，以便让那些看不懂建筑图纸的工匠们按图建造。

然而，不同形式的模仿也同样奏效。尤其是从1997年毕尔巴鄂的古根海姆博物馆建造以来，大型博物馆的一个典型特色就是其空旷的建筑场地，它们通常建在废弃的空地上，其设计目的也异常新奇。但建在复杂场地的小型博物馆也有很多工作要做，对周围环境的智能反应通常比一味的大胆更重要。西班牙巴达霍斯的精品艺术博物馆是由Estudio Arquitectura Hago设计的。它不仅要扮演原有的新古典风格博物馆的扩建结构的角色，还要同那些被历史建筑包围的L形场地争辉。该建筑是面向一个庭院开放的、两座建筑连接而成的一个综合体，其设计用心且风格柔和——充分尊重了周围建筑的高度和立面。建筑内外选择了白色多孔板作为主材料，现代而不奢华。总体来说，设计主要集中在室内，且在外围护结构内建造一系列不同的空间，引进光线，按照建筑师的想法，"该建筑无需组合设计，就会与城市融为一体。"

类似的建筑环境下的更多组合型建筑，便是爱尔兰的沃特福德中世纪博物馆的风格。它由一支来自沃特福德市政厅的团队设计。这个团队现在组成了ROJO工作室来进行设计。这个场地处于一个尴尬的内角区域，面向各种建筑，包括正对面的新古典主义教堂，环境非常复杂。教

Remaining in Australia, the Australian Age of Dinosaurs Museum by Cox Rayner Architects represents a completely different approach. Containing an exhibition of paleontology, and commissioned by a family of ranchers who have given over their land to explore its fossilised relics, this small building is built out over a chasm and appears almost to have grown directly from the rocks, an effect heightened by the fact it has been built from rammed-earth walls excavated from the surrounding soil. Unlike the friendly and colourful library buildings, which set themselves apart from their surroundings to act as focal points for their locale, this museum could hardly be more of its location.

The Dinosaur Museum was built entirely by consultants and contractors who donated their time for free, coming together to work on what they considered to be a worthwhile project. A similar form of collaborative design went into the Museum of Handcraft Paper, in Yunnan Province, China, designed by Trace Architecture Office. This building too, attempts to become completely of its context, in this case by utilising nothing but local materials and techniques, including some of the paper technologies on show in its exhibitions. Like much of the more accomplished architecture coming out of China in recent years, it takes contemporary design tropes (irregular clusters of box-like rooms, frameless glazing, unadorned materials) and juxtaposes them directly with the vernacular, creating a building which wears its crudeness as a virtue. Here, the architects had to create a complex set of physical models in order for the craftsmen, who could not read architectural drawings, to be able to build the scheme at all.

Mimesis can work in other ways as well, however. One of the defining characteristics of the blockbuster museum, especially those built since the Guggenheim Bilbao of 1997, has been the blankness of their sites. Frequently built on empty or derelict land, their architectural purpose is usually to be radically new. But smaller museums on more complicated sites have more work to do, and an intelligent response to their surroundings is often more important than simple boldness. The Fine Arts Museum in Badajoz, Spain, by Estudio Arquitectura Hago, not only has to act as an extension of an existing museum building in a neoclassical style, but has to contend with a L-shaped site hemmed in by historic buildings. The architectural response is an interconnected complex of two buildings opening up to a courtyard, with the design careful and muted – the existing heights and facades of the neighbouring buildings are respected, with a perforated white panel as the primary material choice inside and out, a modern but not exuberant choice. Overall the design is turned to the interior, with a range of

巴达霍斯精品艺术博物馆，西班牙
Fine Arts Museum in Badajoz, Spain

巴达霍斯精品艺术博物馆，西班牙
Fine Arts Museum in Badajoz, Spain

堂的下面是中世纪的古迹，它们是该教堂的一部分，同时也是其展品。建筑师说他们想法是"在强化其历史肌理的特点的同时也要创造一些新的东西"，许多精力已经投入到它的立面设计中——突出的弧形结构，饰面为当地暖色调的石灰石（暗示埋藏在里面的遗迹），加上一系列不规则的带状结构，使该建筑妙趣横生、对比鲜明。这栋建筑暗示了20世纪60年代现代主义所占的分量。入口有一层楼高，用玻璃装饰，打断了立面和悬臂结构的设计。内部的设计将新建筑与腐朽的历史结构结合为统一体。

一方面，这些图书馆和博物馆的设计手法完全不同。图书馆是一种无法确定其公共职能的类型的建筑，通常设计大胆，从它们所在的居民区和商业区中脱颖而出，成为类似于21世纪的会议厅或市政中心。另一方面，博物馆出于一种强烈的辨识意识来开发其建筑手法，无论是从里面手工艺品的特点和材料，还是博物馆开发的当地建筑环境的角度。博物馆以其自身的方式展出大胆的设计，但是却大范围地与环境相适应，而这种差异性至少可以部分由博物馆较少的不确定环境来解释。综合考虑，所有的建筑都会展示当代设计方法运用到小型社区中，并且依然保留其强大的影响力的方法。

different qualities of space and light created within the envelope, with the architects' intention for "the building to be integrated into the city without resorting to a compositional exercise."
More compositional, but dealing with a similar context, is the Medieval Museum in Waterford, Ireland, designed by a team from the Waterford City Council whose are now practicing as ROJO-Studio. The site, again, is an awkward inside corner, facing a variety of buildings including a neoclassical cathedral directly opposite, a situation complicated by the presence of medieval remains underneath which were to become part of the building and its exhibits. The architects state that their intention was to "strengthen the characteristic of the historic tissue while, at the same time, creating something new", and much of the energy has gone into the facade, which is a remarkable curved structure, faced in a warm local limestone (hinting at the ruins embedded within), with a series of irregular bands adding interest and contrast. The building has a hint of the weight of 1960s modernism, with a single storey glazed entrance level undermining the facade and its cantilevers, while the interior revels in the juxtaposition of new construction and decayed historical structures.
On the one hand, the approaches of these libraries and museums are completely different. The strategy for the libraries, a typology which is incredibly unsure of its civic function at this time, frequently attempts to make the buildings bold, standing out from the residential or commercial communities in which they are situated, and becoming something like a 21st century meeting hall or civic centre. The museums, on the other hand, develop their architectural approach from a strong sense of identification, whether it be with the character and materiality of the artefacts, or the local architectural context from which the buildings develop. The museums are in their way equally bold, but in a more intensely contextual fashion, and this difference might be at least partially explained by the less precarious situation for the museum typology at this point. But considered together, all of the buildings demonstrate ways that contemporary design methodologies can be brought down to a smaller community scale, and still retain their power. *Douglas Murphy*

本迪戈图书馆

MGS Architects

改建的本迪戈图书馆（目前是维多利亚第二大图书馆）提供了最先进的图书馆服务，并且设有社区活动和会客空间、静室、咖啡厅、一个志愿者服务中心、一座儿童图书馆以及"游乐小天地"。此外还有艺术品展区，展示了本迪戈的历史及社区艺术项目，并且建筑师为本迪戈地区档案中心阅览室配备了升级的研究设施，此外，大量的互联网接口还增强了其容纳更多社区休闲和学习活动的能力。

这个再开发项目面积约为4000m²，共有两层功能区。甲方希望提供一个更加灵活且包容的图书馆，设计满足了这种的设想，人们可以在这里会面，交流思想，彼此互动。

这个项目将原有的1984年建造的建筑与位于本迪戈历史中心的公共公园的前方结合起来，使之重新成为一种当代的建筑风格和室内设计，实现了公众参与程度的最大化。一个智能建筑管理系统能够跟踪能源的使用情况，并将此信息实时传送给公共场所的管理人，让参观者有机会切实感受到周围能源的使用情况。

这个再开发项目关注三个核心理念，它们结合为一处崭新的环境，拥有本迪戈市民中心的背景，场所的传承价值，公园斜对齐的交界面，以及适当缩放的、直视市政大楼的视野。

增加的脚印设计又将公园重新调整成为一处更加私密的空间。甲方希望图书馆的内部空间环绕着一条内部街道建造，街道又连接着Hargraves大街和Littleton Terrace所在地，而这些增加的设计正是基于此产生的。

通过更好地处理与馆内空间的关系，同时为参观者提供更多随意"浏览"的机会，设计使图书馆前面与公园的氛围更加活跃起来。一间咖啡厅和儿童活动区域令北门前的空间活跃起来。大面积使用的宽大、明亮的落地窗户在图书馆与公园之间产生了透明性，而铰接的、向立面翻转的天篷提供了遮阴。

受原有双高空间的启发，设计师将建筑屋顶向北延伸，并且在两层的上空空间悬挂一个"灯笼"式结构，以在通过内部街道的这段行程内

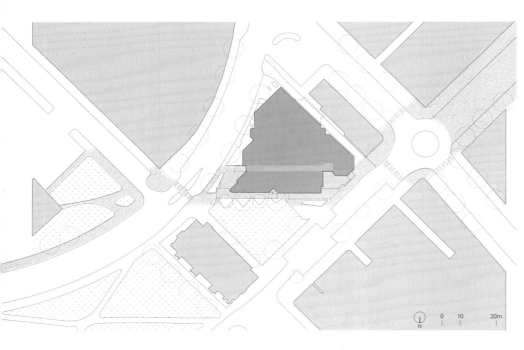

创造一个焦点。这种雕塑般的嵌入结构将新型材料连接在一起,遍布整个翻新项目中,将人们的注意力吸引到了通常很难被注意到的上层空间。

全新的本迪戈图书馆提供了更大的服务范围,成为了社区焦点,在重新开馆的前十天就吸引了20 000人来参观,且会员数量增加了700人。

Bendigo Library

The New Generation Bendigo Library(now the second largest library in Victoria) delivers state of the art library services, community activity and meeting rooms, quiet spaces, a cafe, a Volunteer Resource Center, a children's library and play "cubby", new gallery spaces to showcase Bendigo's history and community art projects, upgraded research facilities for the Bendigo Regional Archives Center Reading Room, and extensive Internet access to augment the community's capacity for leisure and study activities.

The Redevelopment comprises 4,000m² of functional areas over two levels designed to the client's desire to deliver a library that is flexible and inclusive where people can meet, exchange ideas and interact with each other.

The project incorporates the pre-existing 1984 building with frontage to public gardens in the historic center of Bendigo in a contemporary architectural and interior design and responses maximizing the opportunities to engage with the public realm. A Building Management System tracks energy use and transmits this information in "real-time" to the occupants of the public spaces allowing visitors to have the opportunity to interact with this aspect of their immediate surrounds.

The redevelopment is centered on three core ideas which combine to form a new environment: the civic center of Bendigo as a context, the heritage value of the site, the diagonal alignment of the park's interface, and scaling the building appropriately to respond to view lines which direct the eye to the Town Hall building. The additions to the footprint re-scale the park to be a more intimate space and were generated by a desire to order the internal spaces of the library around an internal street linking the Hargraves Street and Littleton Terrace addresses.

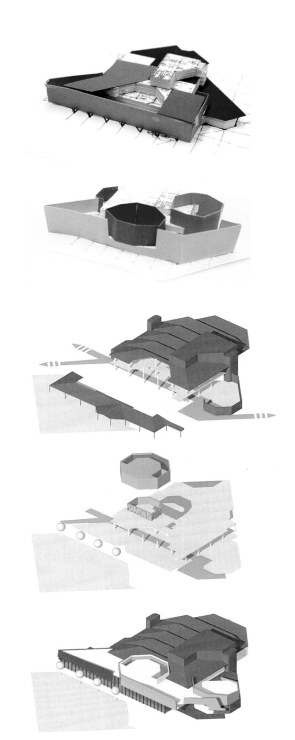

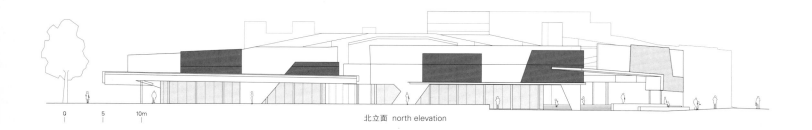

北立面 north elevation

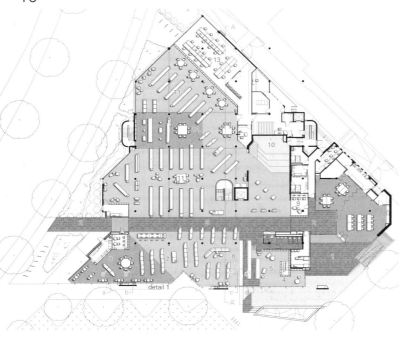

1 入口 2 门厅 3 内部街道 4 文件架 5 儿童室 6 休息室 7 快速挑选区 8 咖啡室
9 志愿者工作室 10 剧院 11 收藏品区 12 游戏区 13 办公室
1. entry 2. foyer 3. internal street 4. cubby hole 5. children's room 6. lounge 7. quick picks
8. cafe 9. volunteers room 10. theater 11. collections 12. game room 13. office
一层 first floor

1 ICT 2 休息室 3 会客室 4 学习区 5 CEO办公室
6 收藏品区 7 阅览区 8 国家档案馆 9 办公室
1. ICT 2. lounge 3. meeting room 4. study area 5. CEO room
6. collections 7. reading room 8. state archive 9. office
二层 second floor

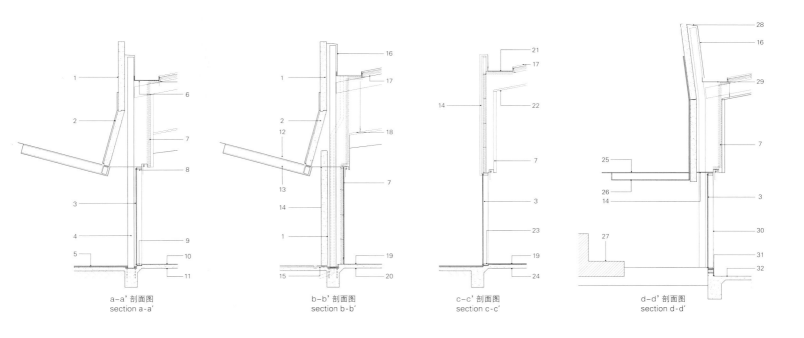

a-a' 剖面图
section a-a'

b-b' 剖面图
section b-b'

c-c' 剖面图
section c-c'

d-d' 剖面图
section d-d'

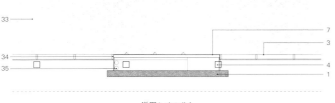

详图1 detail 1

1. precast wall
2. perforated metal sheet fixed to precast wall
3. front glazed anodised aluminium windows
4. column
5. make good to stone paving
6. box gutter and supports laid to falls
7. plasterboard wall lining with insulation
8. pelmet and blind
9. recessed windows sill
10. sika
11. extend new slab and footing
12. canopy top: perforated metal sheet
13. canopy bottom: perforated powder coated metal sheet pan within painted metal
14. pre-finished cladding panel and fixing
15. recessed light fitting with perforated metal cover
16. panel rib cladding to match roof finish
17. metal deck roofing with insulation
18. sculptural plasterboard ceiling
19. carpet
20. extend existing floor slab and provide new footing
21. canopy top: box gutter and supports laid to falls
22. suspended plasterboard ceiling
23. recessed windows sill
24. new slab and footing
25. canopy top: perforated powder coated metal sheet
26. canopy bottom: perforated powder coated metal sheet pan within painted metal structure
27. integrated landscape features
28. flashing fixed and sealed
29. colorbond box gutter on supports
30. window sill flush with external step
31. stone upstand to match ramp finish
32. stone paving
33. downpipes within wall
34. aluminium infill panel to match window finish
35. epoxy flooring

By providing opportunities for a better relationship to the interior and allowing more casual "browsing" opportunities to the visitor the design has improved the activation of the library's frontage with the park. The addition of a cafe and a children's activity area makes the northern frontage an active space. Transparency to the park is achieved with generously proportioned windows shaded by an articulated canopy which lends a twisting movement to the facade.

Taking a cue from the existing double height space the architects extended the roof to the north and hung a "lantern" structure over the two story void to create a focal point in the journey through the internal street. This sculptural intervention ties the new materials used throughout the refurbishment together and brings the eye up to the upper levels which are always more difficult spaces to move into.

The New Generation Bendigo Library provides an expanded scale of services and is a focal point in the community, with enjoying visits by 20,000 people and an increase of 700 members within the first 10 days of re-opening.

项目名称：Bendigo Library / 地点：Bendigo VIC, Australia
建筑师：MGS Architects
建筑设计师：Joshua Wheeler, Eli Giannini
项目经理：Pavan Consulting
改建前建筑的建筑师：Robinson Loo Wyss & Schneider Pty Ltd(1982)
项目建筑师：Chantelle Chiron, Ryan de Winnaar, Rob Compagnino, Kit Kietgumjorn, Babak Kahvazdeh, Gary Yeoh, Sue Buchanan
音效顾问：Marshall Day Acoustics
建筑测量师：Building Issues Pty Ltd
土木/电气/环境/水力/机械顾问：Iriwinconsult
图书馆顾问：KEWS Consulting
景观建筑师：Rush/Wright Associates
甲方：City of Greater Bendigo, Goldfields Library Corporation
有效楼层面积：4,010m² (extension_280m², refurbishment_3,730m²)
造价：USD 9.5m
设计时间：2011.6 / 施工时间：2012.9 / 竣工时间：2014
摄影师：©Andrew Latreille (courtesy of the architect)

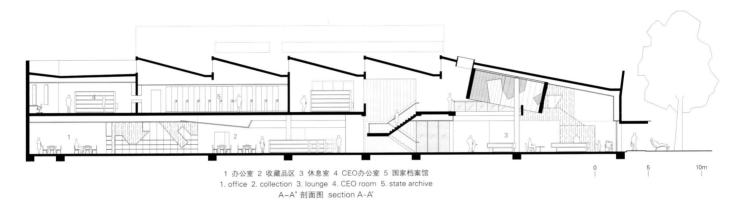

1 办公室 2 收藏品区 3 休息室 4 CEO办公室 5 国家档案馆
1. office 2. collection 3. lounge 4. CEO room 5. state archive
A-A' 剖面图 section A-A'

斯泰普尔顿分馆

Andrew Berman Architect

作为纽约公共图书馆系统的一个新扩建的分馆,其设计方案简明地说就是要将一座独栋建筑与一间式的卡内基图书馆结合在一起。卡内基分馆始建于1907年,由卡雷尔与哈斯丁设计,已服务社区超过100年。该委托项目曾获得彭博社推行的"卓越设计"奖。该奖项是利用一项受邀的竞赛程序,来授予选定的公共委托项目一定的奖励,旨在将纽约城内的最佳设计人才引入公共财政支持的公众建筑领域。

项目要求设计师们为纽约公共图书馆在斯塔顿岛的斯泰普尔顿设计一座经过重新构思设计的、扩建的分馆。毗邻的一块650.3m²的空地划拨给这个项目,用来建造一座1180m²的新图书馆,以便更好地服务于社区,满足当前需要。该社区曾是一片活跃的海港、一些啤酒厂及当地轻工业的旧址。社区的中心Tappen公园曾是维多利亚时期的一个广场,被斯泰普尔顿的市政建筑和商业建筑环绕。当地制造业的长期衰退及滨水区的关闭导致该地区的经济一度停滞。现在,这里居住人口众多,其中多数为移民,他们需要使用图书馆这项资源及公共空间。

新图书馆的设计目标是成为一处富有魅力的、开放且可进入的公共空间,这与纽约公共图书馆的愿景一致,并成为纽约城当地社区的公共资源体系。新图书馆必须是单层设计,将新建筑接合到旧建筑上,并设有残疾人通道。开放式的布局便于员工监控管理,且根据众人意愿将儿童区、青少年区和成人区的空间战略性地分隔开。图书馆应该是开放和吸引人的,要有很强的灵活性,还要结实坚固。建筑师将新建筑建在有自然坡度的地方,这样人们就可以沿坡而不需要爬楼梯进入新街入口了。青少年和成人阅览与研究区域位于新建筑内,中间被一间透明的社区活动室隔开。原有的卡内基图书馆能够从新入口进入,但其修复工作依然忠于原有的设计风格,它的功能是作为儿童阅读室。木质胶合层压板的使用旨在将新建筑打造成为一个高效、给人以温暖氛围且开放的结构。新木材与卡内基图书馆原有的橡木环境和书架之间能够产生对话。使用工厂预制木材可以使结构在限定条件下进行安装,确保材料及工艺的精确度与一致性。现场安装呈流线化,使建筑结构能够更快地安装,并抵挡风雨侵蚀。图书馆通过上方的天窗及玻釉立面透进来的日光进行照明。辐射供暖系统能够对建筑倾斜的混凝土板进行加热。

建筑师与纽约公共图书馆的管理人员及员工密切合作,以期更多地了解他们对于一座现代图书馆的构想。他们还与现有的卡内基图书馆亲密接触,来了解其建筑规模及特性,以期重现这座历史建筑的潜力和方向。建筑师寻求为新图书馆的前方建造一处更加显眼的外部公共空间,期望其与街道对面的Tappen公园——这个维多利亚时代的社会中心产生一种强烈的视觉上和实体上的联系。

项目名称：Stapleton Branch Library
地点：Staten Island, New York
建筑师：Andrew Berman Architect
景观建筑师：Wallace, Roberts & Todd New York
结构工程师：Gilsanz Murray Steficek LLP
MEP工程师：IP Group Engineers
照明顾问：Cline Bettridge Bernstein Lighting Design
地热顾问：Langan Engineering & Environmental Services
甲方：New York Public Library
用地面积：1,811m²
总建筑面积：1,115m²
有效楼层面积：1,115m²
设计时间：2005—2009 / 施工时间：2009—2012 / 竣工时间：2013
摄影师：©Naho Kubota(courtesy of the architect)

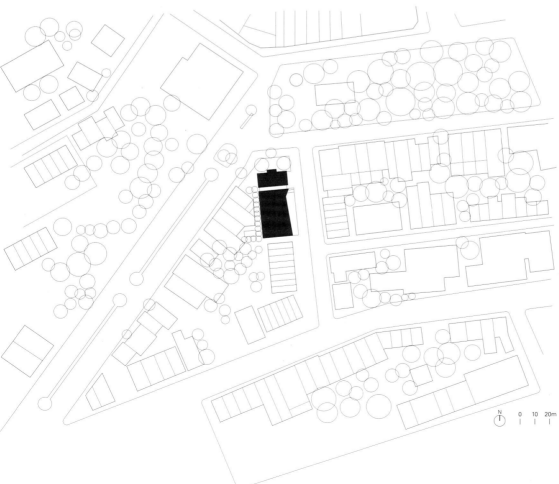

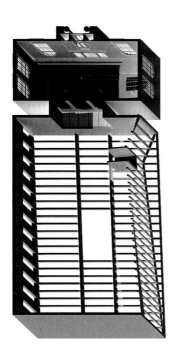

Stapleton Branch Library

The brief for this new and expanded branch library for the New York Public Library system asked for a single story new building to be combined with a single room Carnegie Library. The Carnegie Branch Library from 1907, designed by Carrere and Hastings, had been serving the community for over 100 years. The commission was awarded through a Design Excellence program implemented by the Bloomberg Administration. The program awarded selected public commissions through an invited competition process, with the goal of bringing the best design talent of New York City into the realm of publicly financed civic architecture.

The architects were asked to design a re-conceived and enlarged branch library located in Stapleton, Staten Island for the New York Public Library. An empty adjacent lot was allocated for a 7,000 square foot addition to create a new library of 12,700 square feet that would better serve the community and its current needs. The community was once home to an active seaport, breweries, and local light manufacturing. Tappen Park, the center of the community, was a Victorian era square, ringed by the civic and commercial structures of Stapleton. A long decline in local manufacturing and the closing of the waterfront have led to a period of economic stagnation for the area. Currently, there is a large residential

population, many of whom are immigrants, who use the library as both a resource and a public space.

The new library was intended above all to be an inviting, open, and accessible public space, in keeping with the NYPL vision of itself as a system of public resources for the local communities of New York City. The new library had to be on a single level, stitch new building to old, and be handicapped accessible. An open plan, easily monitored by staff, that provided strategic spatial separations between children's areas, teen area, and adult area was desired. The library had to be open and inviting, flexible, and robust. Working with the sloping grade of the land, the architects sited the new building so that a new street entrance could be accessed from grade, without steps. Teen and adult reading and research areas are located in the new building, separated by a transparent community room. The original Carnegie Library, which is immediately accessed off the new entry, is restored true to its original design, and functions as the children's reading room. Glue laminated timber was used to create an efficient, warm, and exposed structure for the new building. The new timbers speak to the original oak casework and shelving of the Carnegie Library. Use of the shop-fabricated timber allowed for the structure to be fabricated under controlled conditions, ensuring precision and consistency in the material and workmanship. Erection time on site was streamlined, enabling the structure to be erected quicker and made weather-tight. The library is lit by daylight via overhead skylights and through the fritted glazed facades. A radiant heating system efficiently warms the architectural grade concrete slab.

The architects worked closely with the New York Public Library's administration and staff to understand their vision of a contemporary library. They worked closely with the existing Carnegie Library to understand its scale and character, with the desire to revitalize the historic structure's potential and purpose. The architects sought to site the new library to create a clear public exterior space in front of the building, across the street from and with a strong visual and physical connection to Tappen Park, the Victorian center of the original community.

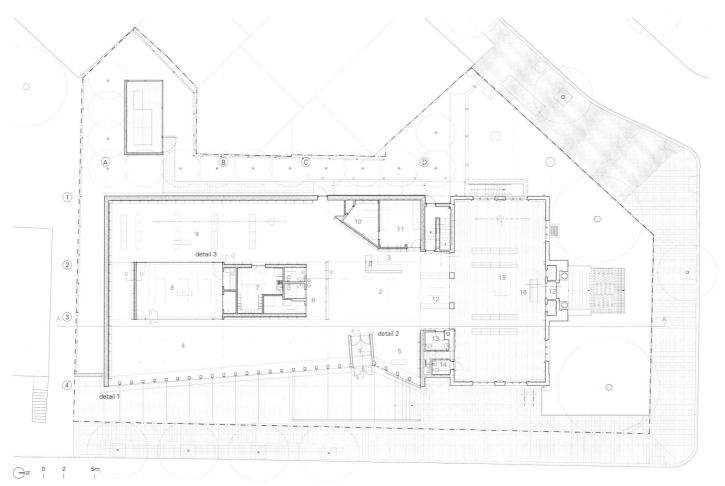

1 入口 2 中央大厅 3 图书借还台 4 青少年区 5 成人阅览区 6 走廊 7 员工休息室 8 社区房间 9 成人区
10 图书管理员办公室 11 员工室 12 前厅 13 员工卫生间 14 卫生间 15 儿童区 16 信息台
1. entry 2. central hall 3. circulation desk 4. teen area 5. adult reading area 6. corridor 7. staff lounge 8. community room 9. adult area
10. librarian's office 11. worker room 12. vestibule 13. staff w.c. 14. w.c. 15. children's area 16. information desk
一层 first floor

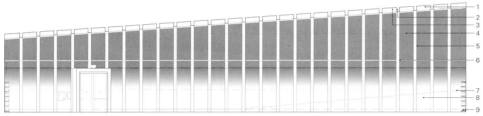

西立面（室内） west elevation (interior)

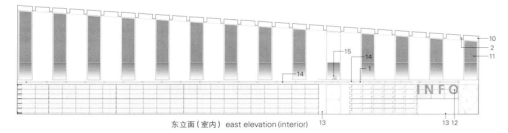

东立面（室内） east elevation (interior)

1. glulam blocking between joists no exposed fasteners
2. glulam joist
3. curtain wall aluminum transom
4. glulam post
5. fritted insulated glazing unit 50% frit coverage
6. aluminum tube at transom
7. aluminum glazing channel
8. grade beyond
9. bookcase
10. glulam girder
11. fixed window with graded frit on #2 surface of insulated glazing unit
12. graphic on wall, behind open-backed bookcase
13. fixed steel bookcase with back panel
14. bookcase light fixture
15. exit light on soffit

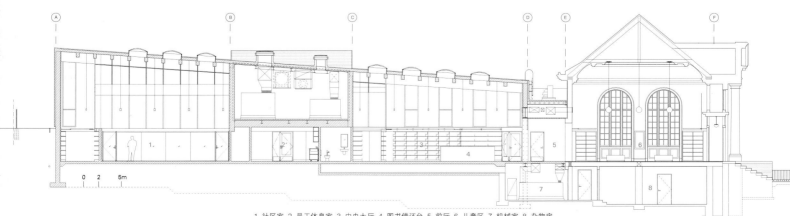

1 社区室 2 员工休息室 3 中央大厅 4 图书借还台 5 前厅 6 儿童区 7 机械室 8 杂物房
1. community room 2. staff lounge 3. central hall 4. circulation desk 5. vestibule 6. children's area 7. mechanical 8. utility room
A-A' 剖面图 section A-A'

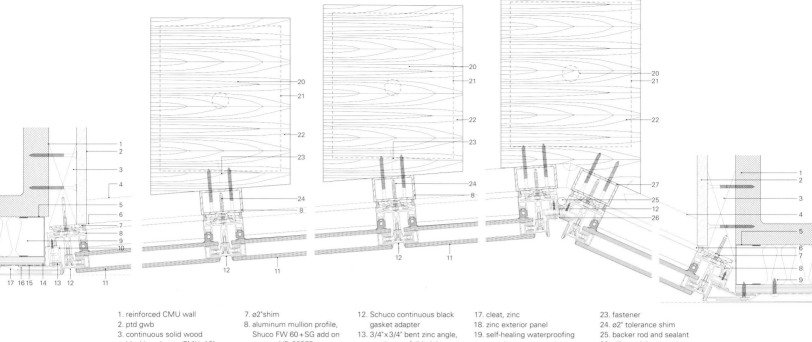

1. reinforced CMU wall
2. ptd gwb
3. continuous solid wood blocking shot to CMU, 16" o.c.
4. 2 1/2"×1 1/2"×1/8" aluminum tube below
5. 2 1/2" metal framing
6. paintable silicone
7. ø2"shim
8. aluminum mullion profile, Shuco FW 60+SG add on system V8-99375
9. batt insulation
10. Schucco PVC block
11. insulated glazing unit with rebated spacer
12. Schuco continuous black gasket adapter
13. 3/4"×3/4" bent zinc angle, continuous full height
14. 1/2" plywood
15. building felt
16. bent zinc flashing, continous full height
17. cleat, zinc
18. zinc exterior panel
19. self-healing waterproofing membrane
20. 1"ø bolt
21. 7 1/4"× 9" base plate
22. douglas fir glulaminated post, shop bevel face
23. fastener
24. ø2" tolerance shim
25. backer rod and sealant
26. silicone joint
27. 2 1/2"×1 1/2"×1/8" aluminum tube spacer

详图1 detail 1

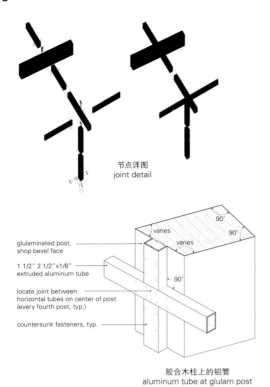

节点详图 joint detail

- glulaminated post, shop bevel face
- 1 1/2" 2 1/2"x1/8" extruded aluminum tube
- locate joint between horizontal tubes on center of post (every fourth post, typ.)
- countersunk fasteners, typ.

胶合木柱上的铝管 aluminum tube at glulam post

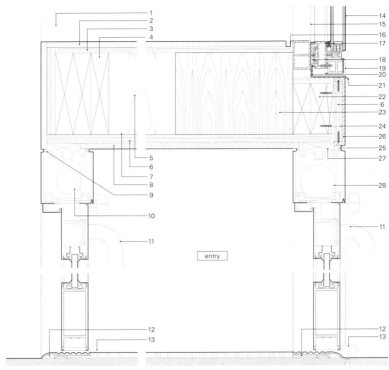

a-a' 剖面图 section a-a'

1. specified exit motion sensor
2. 1/2" mdf painted, concealed fasteners
3. 1/2" plywood
4. 2"x8" wood framing
5. SAB insulation
6. 3/4" plywood
7. 7 1/4" light gage mtl framing 16"o.c.
8. 3/4"x5 1/2" cvg douglas fir t+g boards, blind nailed
9. paintable silicone
10. concealed overhead door closer
11. specified door pull
12. specified saddle aligns edge with door frame
13. specified offset pivot hinge, recessed floor mounted plate
14. insulated glazing unit with rebated spacer
15. 2 1/5"x1 1/2"x1/8" aluminum tube alloy 6063-T52
16. paint out reveal prior to installation of mdf
17. aluminum transom profile, Schuco FW60+SG add on system V8-99375
18. ø2" tolerance shim, max 3/8" thick of suitable hardness
19. Schuco PVC block wrapped in zinc, no exposed fasteners
20. continuous shelf-healing membrane waterproofing
21. zinc flashing with drip edge bond braking tape between zinc & aluminum
22. wood blocking
23. 10"W x 7 1/4"D glulam beam
24. concealed fasteners
25. backer rod and sealant
26. 6"x1.92" aluminum channel, no exposed fasteners, channel flanges to be cut to 1" length
27. blocking and insulation as required
28. specified power assist door closer
29. 2"x6" light gage metal framing
30. 2"x6" wood studs
31. 1/2" douglas fir t+g boards, blind nailed
32. glulam douglas fir post
33. paintable silicone
34. door frame
35. 1 1/2"x4"x1/8" aluminum tube

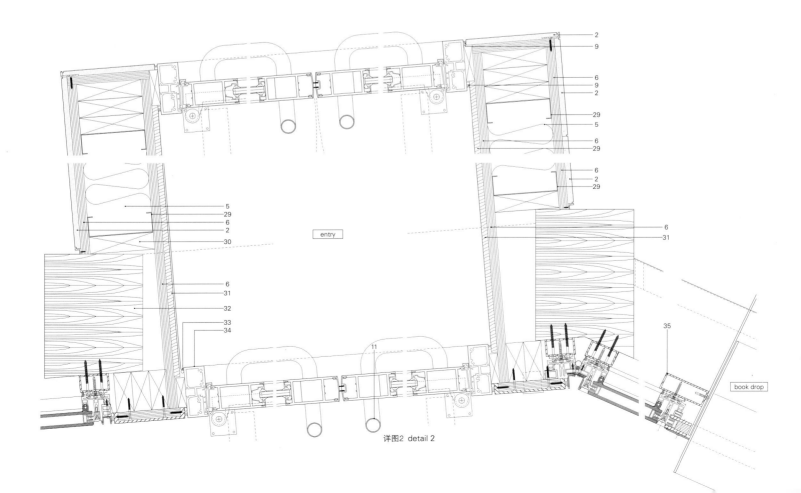

详图2 detail 2

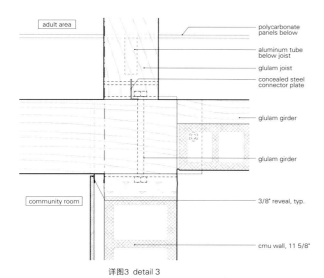

详图3 detail 3

1. ceiling finish type C-1
2. glulam joist
3. glulam girder
4. fasteners
5. duct hanger strapping 1" wide typ.
6. supply air duct inside ø=27"
7. air diffuser
8. c.o.duct el. 16'-8"
9. full height polycarbonate sheet
10. aluminum post
11. light fixture
12. aluminum plate
13. aluminum beam
14. cmu wall
15. bookshelf, 12 1/2" deep with backpanel
16. finish concrete slab
17. structural slab
18. base B-4
19. architectural glass wall
20. fire damper

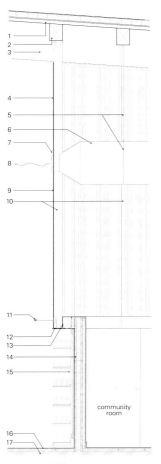

b-b' 剖面图 section b-b'

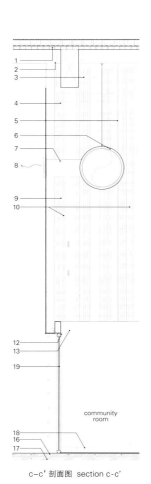

c-c' 剖面图 section c-c'

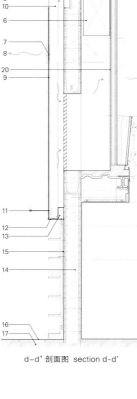

d-d' 剖面图 section d-d'

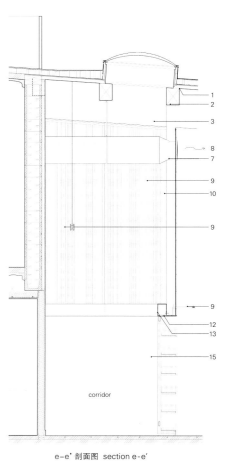

e-e' 剖面图 section e-e'

Vennesla图书馆和文化中心

Arkitektfirma Helen & Hard AS

挪威Vennesla市的新图书馆项目包括一座图书馆、一座咖啡馆、会客室和行政区，并与已有的社区活动室和学习中心连接在一起。

为了打造一处好客的公共空间，建筑师将所有的主要公共职能集中到一处宽敞的空间中，允许人们无论从内部或外部都可清晰地看见与家具结合的建筑结构，以及多元空间的交界面。一条整体通道将城市生活也纳入其中。此外，项目书要求新建筑面向主要城市广场开放并且方便由此进入，与现有城市肌理紧密结合。这是通过采用室外的大型玻璃立面和凉廊（提供了一处免受风吹雨淋的室外座位区）而实现的。

在这个项目中，建筑师开发了一种肋状结构理念，创造了非常实用的综合结构，它们将木结构与所有技术设备和室内结合起来。

27根由预制胶合层压木构件和数控切割胶合板制成的木肋构成了整座图书馆。这些木肋形成了屋顶的几何轮廓，也形成了内部宽敞且开放的空间的起伏方向，开放空间内还设有私人学习空间，嵌在周边区域。

每根木肋都包含一个胶合层压木梁柱、含有声学吸收剂的空调管道、作为灯罩和闪光标志的弯曲玻璃窗格，以及一体化的阅览座位和书架。

渐次改变形状的肋状结构，是顺应相邻两座建筑物的形状而产生的，同时也因图书馆的不同隔间对空间的品质及功能的需求各异而产生。两端的立面都依照场地的特定需求设计。在主入口，肋状结构形成了凉廊，横跨整个广场。建筑沿自然场地的线条面向西／南方向进行布局，并根据内部设计方案及高度要求向街道弯折下来。这一侧的立面安装了固定的垂直遮阳板。遮阳板将建筑聚拢在一起，使建筑相对于两侧相邻建筑显得格外突出。

设计的主要意图是通过"填实空隙"的概念以及将高标准节能解决方案应用于所有新部件的方式来降低三座建筑的能源需求。这座低能耗图书馆在挪威的能源使用评估体系中被评定为A级。建筑师旨在最大限度地利用木材。仅建筑本体就使用了总计超过450m³的胶合木板。所

有肋状结构、内外墙体、电梯井、板材和部分屋顶都是由胶合木板制成的。所有的胶合木板都至少暴露在一侧，或两侧均有。

一个将结构、技术基础设施、家具与内部空间融合成为一个建筑元素的整体结构，创造了一种强烈的空间特性，满足了甲方想要标示出城市文化中心的初衷。

Vennesla Library and Culture House

The new library in Vennesla comprises a library, a cafe, meeting places and administrative areas, and links an existing community house and learning center together.

Supporting the idea of an inviting public space, all main public functions have been gathered into one generous space allowing the structure combined with furniture and multiple spatial interfaces to be visible in the interior and from the exterior. An integrated passage brings the city life into and through the building. Furthermore, the brief called for the new building to be open and easily accessible from the main city square, knitting together the existing urban fabric. This was achieved using a large glass facade and urban loggia providing a protected outdoor seating area.

In this project, the architects developed a rib concept to create usable hybrid structures that combine a timber construction with all technical devices and the interior.

The whole library consists of 27 ribs made of prefabricated glue-laminated timber elements and CNC-cut plywood boards. These ribs inform the geometry of the roof, as well as the undulating orientation of the generous open space, with personal study zones nestled along the perimeter.

原有的场地，位于一座老电影院和一座文化学校之间，面向城镇广场
the existing site, between an old cinema and a cultural school, facing the town square

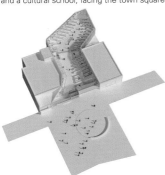

原有的电影院门厅进行了切割，使新建筑和城镇广场之间的交界面变得最大
the existing cinema foyer is cut to maximize the interface between the new building and the town square

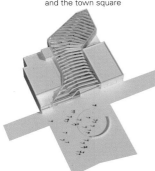

建筑环境中的肋状结构
rib structure in context

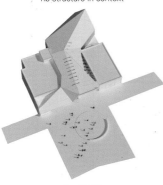

竣工的建筑
the complete building

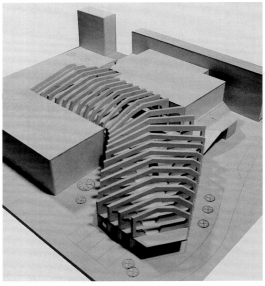

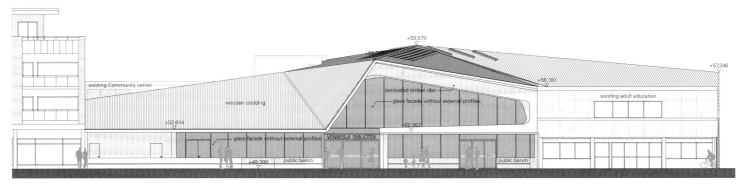
东立面 east elevation

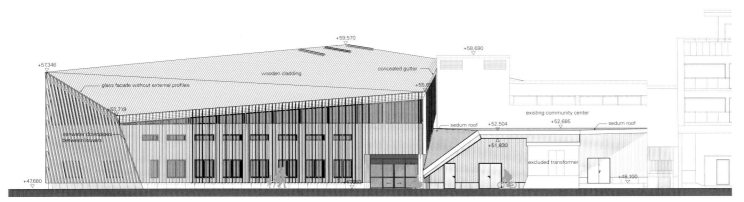
西立面 west elevation

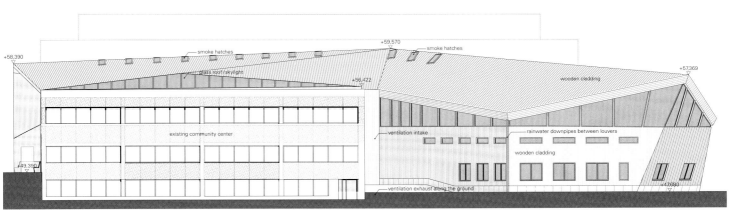
北立面 north elevation

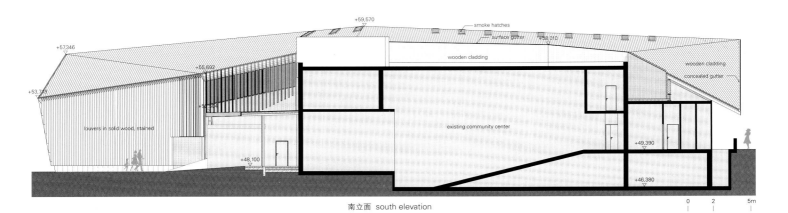
南立面 south elevation

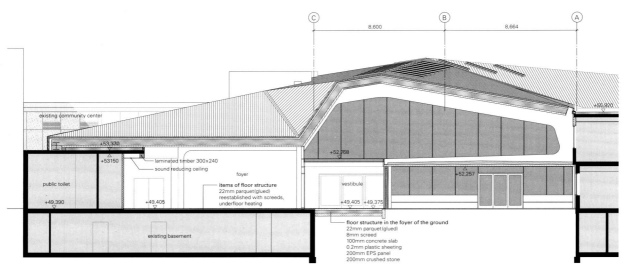

A-A' 剖面图 section A-A'

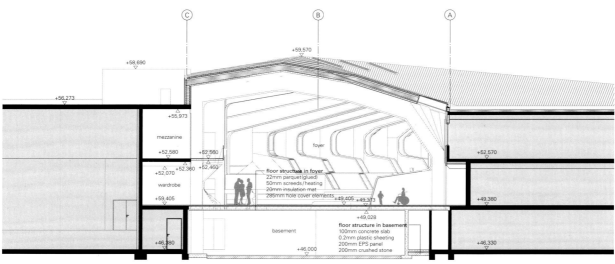

B-B' 剖面图 section B-B'

C-C' 剖面图 section C-C'

Each rib consists of a glue laminated timber beam and column, the air conditioning ducts which contain acoustic absorbents, bent glass panes that serve as lighting covers and signs, and integrated reading niches and shelves.

The gradually shifting shapes of the ribs are generated through adapting to the two adjacent buildings and also through spatial quality and functional demands for the different compartments of the library. Each end facade has been shaped according to the specific requirements of the site. At the main entrance, the rib forms the loggia which spans the width of the entire square. Against south/west the building traces the natural site lines, and the building folds down towards the street according to the interior plan and height requirements. On this side, the facade is fitted with fixed vertical sunshading. This shading also gathers the building into one volume which clearly appears between the two neighboring buildings.

A main intention has also been used to reduce the energy need for all three buildings through the infill concept and the use of high standard energy saving solutions in all new parts. The library is a "low-energy" building, defined as class "A" in the Norwegian energy-use definition system. The architects aimed to maximize the use of wood in the building. In total, over 450m³ of glulam wood has been used for the construction alone. All ribs, inner and outer walls, elevator shaft, slabs, and partially roof, are made in glulam wood. All glulam is exposed on one or both sides.

A symbiosis of structure, technical infrastructure, furniture and interior in one architectonic element creates a strong spatial identity that meets the client's original intent to mark the city's cultural center.

项目名称：Vennesla Library and Culture House
地点：Vennesla, Norway
建筑师：Arkitektfirma Helen & Hard AS
项目团队：Reinhard Kropf, Siv Helene Stangeland, Håkon Minnesjord Solheim, Randi Augenstein
有效楼层面积：1,938m² / 造价：NOK 66,400,000
竞赛时间：2009 / 竣工时间：2011
摄影师：©Emile Ashley-p.32~33, p.39bottom, p.44~45
©Hufton+Crow-p.34~35, p.39top, p.41

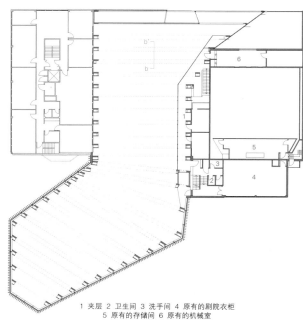

1 夹层 2 卫生间 3 洗手间 4 原有的剧院衣柜
5 原有的存储间 6 原有的机械室
1. mezzanine 2. toilets 3. washroom 4. existing theater wardrobe
5. existing storage 6. existing mechanical room
夹层 mezzanine floor

1 技术间 2 休息/会议/学习室 3 本地历史收藏品室 4 说书室 5 教室
6 办公室 7 存储室 8 卫生间 9 低层门厅 10 衣柜 11 清洁室 12 杂志区
1. technical room 2. break/meeting/study room 3. local history collection
4. storytelling room 5. classroom 6. offices 7. storage 8. toilets
9. lower foyer 10. wardrobe 11. cleaning room 12. magazines
地下一层 first floor below ground

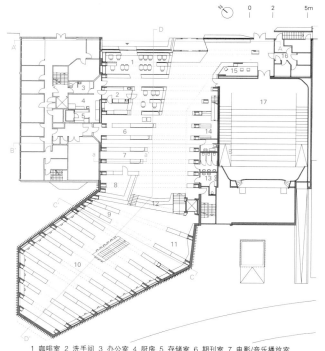

1 咖啡室 2 洗手间 3 办公室 4 厨房 5 存储室 6 期刊室 7 电影/音乐播放室
8 图书馆接待处 9 儿童收藏室 10 图书馆 11 青少年收藏室 12 剧场 13 卫生间
14 衣柜 15 影院接待处 16 原有的卫生间 17 原有的大厅
1. cafe 2. washroom 3. office 4. kitchen 5. storage 6. periodicals 7. films/music
8. library reception 9. children collection 10. library 11. youth collection 12. amphitheater
13. toilets 14. wardrobe 15. cinema reception 16. existing toilets 17. existing hall
一层 ground floor

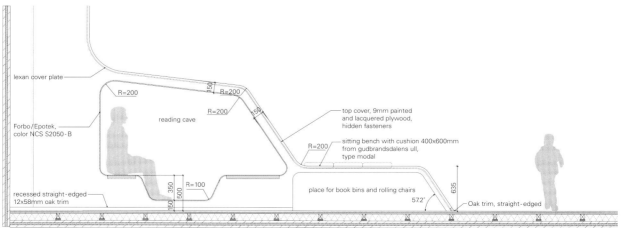

a-a' 剖面图 section a-a'

42

⑭ ⑮ ⑯ ⑰ ⑱ ⑲ ⑳ ㉑ ㉒ ㉓ ㉔ ㉕ ㉖

non-hatch

concealed gutter

library

reading corner

wooden board for shading

floor structure in library
22mm parquet(glued)
22mm chipboard
73x48mm joists and 100mm insulation
27mm rubber pads
100mm solid

floor structure library of meeting
22mm parquet(glued)
22mm chipboard
73x48mm joists and 100mm insulation
27mm rubber pads
100mm solid

floor structure library of technical room
22mm wooden(glued)
22mm chipboard
felt paper
200mm hole cover elements

+50,626   +50,770   +50,726
+50,526   +50,526

storytelling room   local history collection   break / meeting / study room

ventilation air extraction behind telescope amphitheater

utility room   cleaning

+47,655

floor structure of culvert
22mm parquet (glued)
28mm screed
200mm concrete slab

floor structure reinforced edge
22mm parquet(glued)
8mm screed
200mm plastic wrap
100mm EPS panel
100mm crushed stone

floor structure in basement
22mm parquet(glued)
8mm screed
100mm concrete cover
0.2mm plastic wrap
200mm crushed stone

+47,630   +47,680   +47,650

culverts   +47,430

floor structure of culvert
250mm concrete slab
0.2mm plastic wrap
100mm XPS

+46,380

floor structure in basement
100mm concrete slab
0.2mm plastic sheeting
200mm EPS panel
200mm crushed stone

+46,000   +46,380

+45,680

D-D' 剖面图 section D-D'

roof cladding, heartwood pine(23x98mm top board, 23x69mm bottom board)
36/48mm batten, heartwood pine
corrugated steel roofing
batten
sloping batten
50mm continuous non-combustable insulation layer over the ribs
300mm rafters with insulation
50mm insulation
ceiling cladding, varied dimension
glulam rib

add-on, 50mm painted birch plywood

50mm add-on

glulam rib

bookshelves   study place between ribs

supply air diffused from the bottom of the bookshelves

详图1 detail 1

0   1   2m

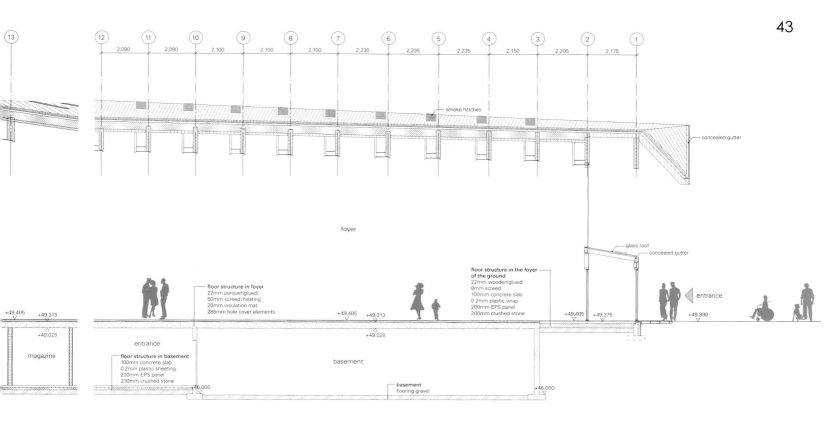

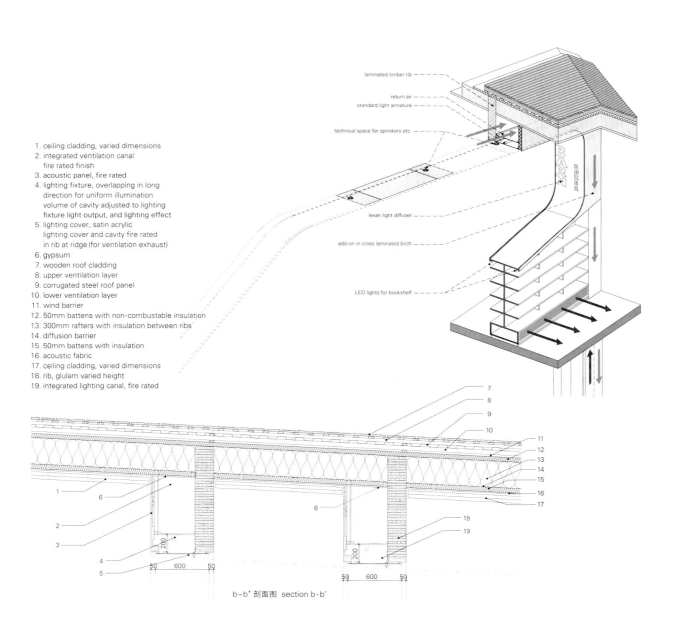

b-b' 剖面图 section b-b'

Cooroy图书馆
Brewster Hjorth Architects

Cooroy图书馆和数字信息中心是Cooroy文化专区（Mill Place Cultural Precinct for Cooroy）总体规划中的一部分，Cooroy是一座充满活力的乡村小镇，位于澳大利亚阳光海岸的内陆地区。

建筑坐落于原有的、画廊与北部绿地之间的斜坡上。依斜坡而建的两座展馆之间保持的横向落差为5~6m，而建筑切入到其中，呈弯曲状，方便连通公园和小镇。西侧的展馆将上部的公共空间延续到能俯瞰公园的草皮屋顶中，而屋顶的曲线形态形成了一个荫蔽的后庭院，为内部空间带来了穿堂风。建筑的ESD系统是这一理念的核心。其借助于地面提供的荫蔽的位置优势，来驱动建筑的空气冷却和通风系统，同时为免受日间温度波动产生的影响提供了强有力的保温性能。

新建筑在小镇与文化区之间建立了一个非常重要的连接，有助于为画廊和艺术工作室提供一个焦点。它的草皮屋顶、封闭游廊和室外庭院都是重要的社区展示、互动空间。

面向北侧深度弯曲的游廊延续了主要大街的人行道，且连接了公共入口。游廊的格调和规模创造了一个与小镇的零售区的视觉上和情感上的连接。建筑的两座展馆反向弯曲，遵循了斜坡的轮廓而建造。上部的草坪公共区域为曲线的中心，延伸至西侧展馆的屋顶之上，重塑了原有的山坡地貌。委托建造的大型雕塑标出了建筑的交叉点/轴心，而游廊尽头一侧设置了一个全新的多面灯笼，作为雕塑的对应物和新建筑的标记。

建筑包括一座大型的公共图书馆，带有数字培训室、社区活动室、社区休息室和记载当地传统的遗产室。图书馆的设计旨在面向不同群体：儿童、青年、学生、IT用户和老年社会群体。图书馆中的单独区域有着醒目的"舱式"外观设计，坐落于建筑下部，与建筑结构分离。

景观和生态建筑工程解决方案完全融入了设计中。利用创造性的结构工程来支持绿色屋顶的设计是成功的关键所在。

这座由国家资金资助而建造的建筑的设计方案被限定在非常严格的预算之内。设计过程包括成本规划和价值管理。地面遮阴结构利用了稳定的地下温度来被动控制内部舒适度。混合模式/置换通风系统使用的是来自地底的预先冷却的空气，因此能够达到显著的节能效果。建筑内部能自行发电，还会收集雨水，以循环利用，进行冲厕及灌溉。

整个阶段，经过设计师们与社区反复磋商，项目才得以完成。这座建筑非常受人喜爱，并且利用率颇高。

Cooroy Library

The Cooroy Library and Digital Information Hub was developed as part of Mill Place Cultural Precinct for Cooroy, a vibrant rural town located in the hinterlands of the Sunshine Coast.

The building sits into the existing slope between the gallery and northern parklands. It cuts into the 5~6m cross-fall arranged in 2 pavilions aligned across the slope; the building is curved so that it addresses the park and the township. The western pavilion continues the upper level public space onto its grassed roof which looks out over the park, while its curve creates a sheltered rear courtyard allowing cross ventilation to the interior spaces. The building's ESD systems are core to its concept. They rely on its earth sheltered location to drive its air cooling and ventilation systems as well as providing strong insulation from diurnal temperature fluctuations.

The new building forms an important link between the town and

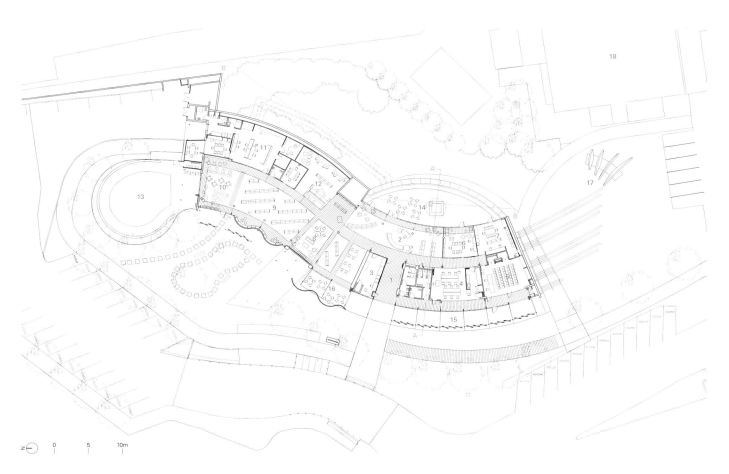

1 入口 2 顾客休息区 3 还书区 4 连接中心 5 社区功能间 6 休息室 7 遗产研究室 8 多媒体休息室 9 收藏品室
10 儿童图书馆 11 员工工作区 12 青少年休息室 13 室外活动区 14 室外下沉庭院 15 游廊 16 咖啡阅览室 17 雕塑花园 18 艺术画廊
1. entry 2. customers 3. returns 4. connection hub 5. community program room 6. lounge 7. heritage study 8. multimedia lounge 9. collection
10. children's library 11. staff work area 12. youth lounge 13. external activity area 14. external sunken courtyard 15. veranda 16. cafe reading 17. sculpture garden 18. art gallery
一层 first floor

cultural precinct, helping to provide a focus for the gallery and art-workshop areas. Its grassed rooftop, wrapping veranda and outdoor courtyards are important community display and interaction spaces.

The deep curved north-facing veranda continues the main street footpath connecting to the public entry. The rhythm and scale of the veranda create a visual and emotional link with the town's retail strip. The building's two opposing curved pavilions follow the contours of the slope. The upper grassed public area which forms the center of the curve extends onto the roof of the western pavilion, recreating the original hillside. The junction/pivot point is marked by the large commissioned sculpture and the veranda terminates at the new faceted lantern which acts as a counterpoint to the sculpture and a marker for the new building.

The building includes a large new public library with digital training rooms, community rooms, community lounges, and heritage rooms for work on local records. The library has been designed with a focus on disparate groups: children, youth, students, IT users and the older community. Individual areas in the library are defined by strikingly designed "pods" sitting below and separate from the building structure.

Landscape and sustainable building engineering solutions have been fully integrated in the design. Creative structural engineering to support the green roof was key to success.

The state funded building was designed to a very strict budget. The design process included cost planning and value management. The earth sheltered structure takes advantage of stable underground temperatures to passively control internal comfort. The mixed mode/displacement air system uses pre-cooled air from the underground labyrinth, giving significant energy savings. The building generates its own electricity and harvests rainwater for recycling used in toilets and irrigation.

The program was developed closely with continuing consultation between the designers, and the community through all stages. The building is much loved and highly used.

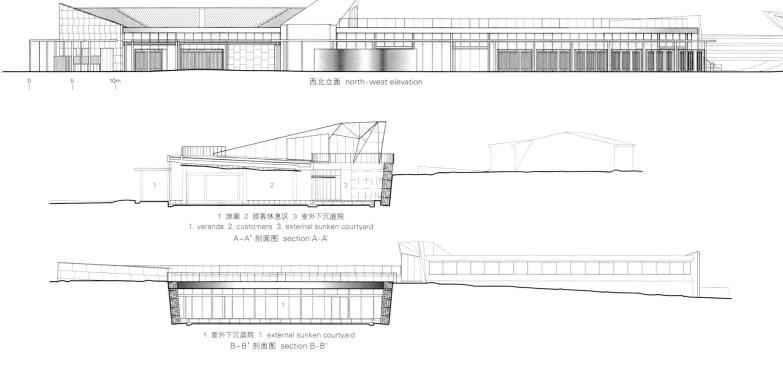

西北立面 north-west elevation

1 游廊 2 顾客休息区 3 室外下沉庭院
1. veranda 2. customers 3. external sunken courtyard
A-A' 剖面图 section A-A'

1 室外下沉庭院 1. external sunken courtyard
B-B' 剖面图 section B-B'

项目名称：Cooroy Library / 地点：Sunshine Coast, Queensland, Australia
建筑师：Brewster Hjorth Architects
设计建筑师：Ian Brewster / 项目经理：Luigi Staiano
项目团队：architectural staff _ Maria Colella, Thuy Hang Nghiem, Siu Wang, Carnie Chu / local architects _ Ken Robinson, Jolyon Robinson
结构顾问：Robert Bird Group / 水力顾问：Medland Metropolis
机械/电气/环境顾问：Steenson Varming Pty Ltd
景观顾问：Place Design Group / 造价顾问：Graham Lukins Partnership
音效顾问：Acoustic Logic Consultancy / 交通顾问：PB Traffic Engineering
BCA：Vic Lilli and Partners
可达性设计顾问：Morris Goding Accessibility Consultants
规划：Martoo Consultant
制图：Zinc
建筑商：Hutchinson Builders
主管：Peter Thomson
用地面积：6,280m² / 总建筑面积：1,350m² / 有效楼层面积：1,350m²
设计时间：2010 / 施工时间：2010 / 竣工时间：2010
摄影师：
©Paul Jackson (courtesy of the architect) - p.46, p.48, p.49[bottom], p.51, p.52[middle], 53[top]
©Christian Mushenko (courtesy of the architect) - p.49[top], p.50, p.52[top, bottom], 53[bottom]

巴达霍斯精品艺术博物馆扩建项目的主要建筑理念是为了重新获得一个身份：一处全新的、通过文化内容（博物馆）能与城市环境（城市）互动的建成环境（建筑）。建筑商要让"最美丽的建筑"展览艾斯特雷马杜拉的伟大艺术家，如苏巴朗、内罗周或者卡瓦斯等人的大量收藏品。

建筑师建造了一座综合性建筑，其起点是对原有博物馆（位于巴达霍斯市中心的一座列管建筑中）进行扩建。这座综合建筑包括两座全新的、通过庭院连接的建筑，它们面向城市的两条不同的街道开放。由于这个地方汇集了较难处理的环境要素（考古遗迹、城墙，以及列管建筑的修复），所以这里需要一个有力且连续的建筑形式。

两个建筑体块应不同的需求进行设计。一个体块里面陈列着永久性收藏品，体块依附于原有的结构，使新楼层对于原有的建筑来说也非常适应。另一个体块则用于举办临时的展览，四楼是总部所在地。两个体块之间通过一个庭院相连，人们可以对这座综合性建筑进行多种方式的利用。夏天还有可能举行群众爵士音乐会。

一个倒写的"L"形地理规划满足了建筑的功能、空间、结构和科技方面的需求。这个规划通过覆层、预应力穿孔混凝土板（混凝土板的规格和图案看起来就像是在建筑的表层印制信息）得以实现。

带白色边框的大型玻璃窗格呈离心状，嵌入带有斜边的金属覆层周围，形成了每座建筑的入口处。面向玻璃的双高大厅的后面是一个穿孔的金属平台，平台设有纤细的白色栏杆，形成了永久性展览空间的一层。在三层，地板大角度地向上倾斜，形成了斜切的立面。白色光滑的地板上镶嵌的长条玻璃使人们从下面和上面都可以看到画廊。建筑新建部分的楼层通过一段狭窄的白色楼梯连接，且由于头上的圆形天窗而让楼梯充满了自然光。在临时展览空间的上层，室内的窗户设有黑色的框架，人们的视野从新空间一直延伸到带有华丽金属装饰的赤褐色楼梯上。

巴达霍斯精品艺术博物馆
Estudio Arquitectura Hago

东南立面 south-east elevation

由此产生的高效但不张扬的结果允许建筑融入到城市中，且超越了建筑的物理界限，将规划的空间变成城市环境内真正的闪耀明星。这个项目展现了其与美学的撞击，记忆与力量及时地留存其中，并将整座博物馆变成了巴达霍斯市中心的一个全新的社会、城市和艺术实体。

Badajoz Fine Arts Museum

The backbone of the architectural strategy for the extension project of the Fine Arts Museum in Badajoz is meant to regain an identity: a new built environment(architecture) that interacts with the urban context(city) through its cultural content(museum). The architects would like to make "the most beautiful building" to house the huge collection of Extremadura's great artists, like Zurbarán, Naranjo or Covarsí.

The architects project a complex whose starting point is the expansion of the existing museum, located in a listed building in the center of Badajoz. The complex includes two new buildings that are connected through a courtyard and are opened to two different streets of the city. Due to the difficult circumstances that come together in this place(archaeological remains, party walls and the rehabilitation of the listed building) a powerful and coherent architectural response is required.

Two blocks are designed for different requirements. One building

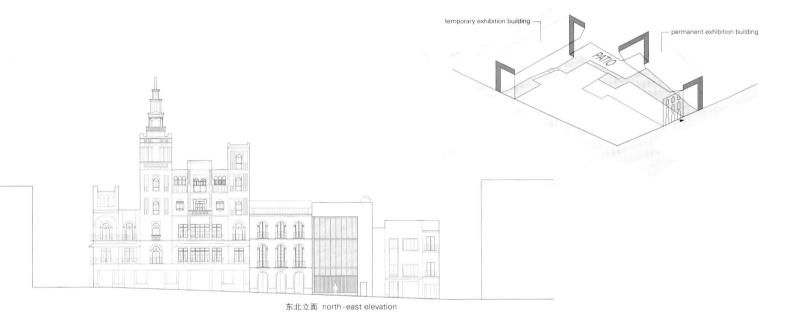

东北立面 north-east elevation

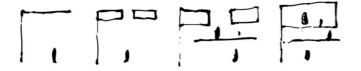

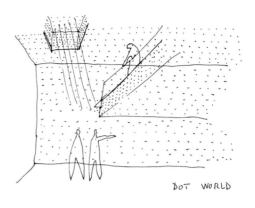

项目名称：Fine Arts Museum
地点：Badajoz, Spain
建筑师：Antonio Álvarez-Cienfuegos, Rubio & Emilio Delgado-Martos
合作者：Ignacio Herreros, Javier Bachiller & Iago Sánchez
结构工程师：Andrés Rubio Morán, Juan Ruiz y Eliseo Pérez
设备工程师：Carlos Úrculo, Úrculo Ingenieros
用地面积：1,435.57m²
总建筑面积：3,298.39m²
有效楼层面积：2,650.69m²
竞赛时间：2007
施工时间：2011—2014
摄影师：©Fernando Alda (courtesy of the architect)

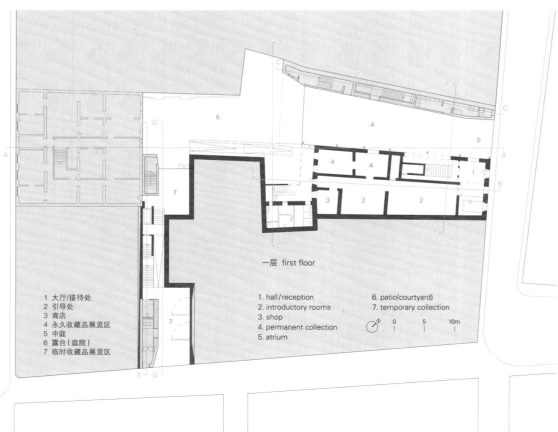

一层 first floor

1. 大厅/接待处
2. 引导处
3. 商店
4. 永久收藏品展览区
5. 中庭
6. 露台(庭院)
7. 临时收藏品展览区

1. hall/reception
2. introductory rooms
3. shop
4. permanent collection
5. atrium
6. patio(courtyard)
7. temporary collection

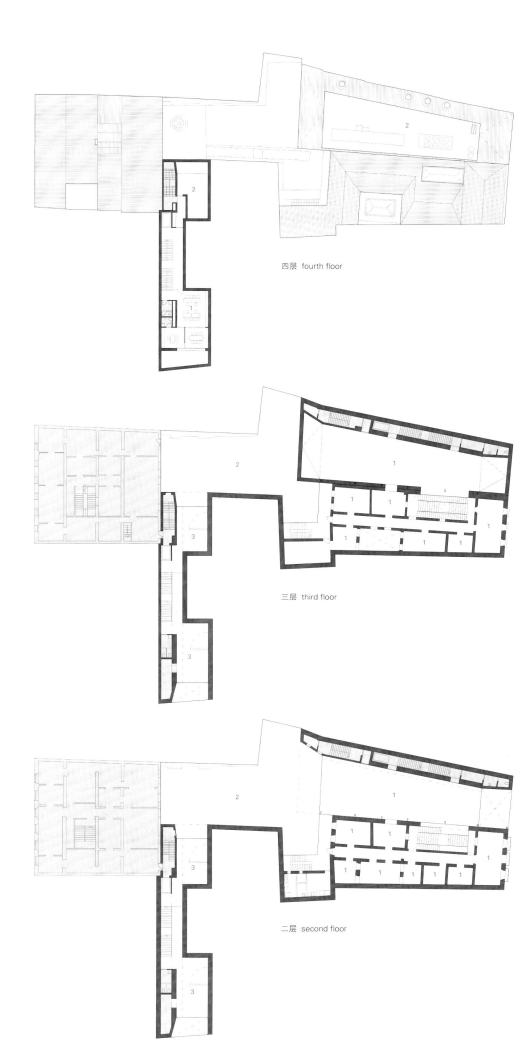

1 办公室
2 设备区

1. offices
2. facilities area

四层 fourth floor

1 永久收藏品展览区
2 庭院
3 临时收藏品展览区

1. permanent collection
2. courtyard
3. temporary collection

三层 third floor

1 永久收藏品展览区
2 庭院
3 临时收藏品展览区

1. permanent collection
2. courtyard
3. temporary collection

二层 second floor

houses a permanent collection and it is attached to the original construction, adapting the new levels to the existing ones. The other block can be used for temporary exhibitions and it houses the headquarters offices on the fourth floor. They are connected by a courtyard, allowing to use the complex in different ways. In summer it is possible to organize jazz crowded concerts.

A typological scheme organized by an inverted "L", responds to the functional, spatial, structural and technical aspects. This scheme is fostered(inside and outside) through a cladding, a perforated pre-stressed concrete panel, whose scale and pattern seem to print a message on the skin of the building.

Large panels of glazing with white frames are set off-center into beveled metal-clad surrounds to form the entrance points to each of the buildings. Behind a double-height lobby faced with glass, a perforated metal platform with slim white railings forms the first floor of the permanent exhibition space. On the third level, the floor slopes dramatically upwards to account for the beveled facade. Strips of glazing cut through the glossy white floors and offer views into the galleries below and above. Levels in the new parts of the building are connected by a narrow white staircase, which is filled with natural light thanks to circular skylights overhead. On the upper floors of the temporary exhibition space, interior windows with deep black frames look from the new space into a terracotta-toned stairwell with ornate metalwork.

The efficient and silent result allows the building to be integrated into the city, overflowing the physical limits of the buildings and turning the proposed spaces into real stars of the urban environment. This project represents an encounter with beauty where the memory and its power remain in time, and convert the whole museum into a new social, urban and artistic reality in the center of Badajoz.

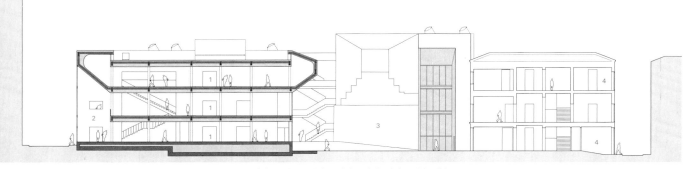

1 永久收藏品展览区 2 中庭 3 庭院 4 收藏品存储区&修复工作间
1. permanent collection 2. atrium 3. courtyard 4. collection storage & restoration workshop
A-A' 剖面图 section A-A'

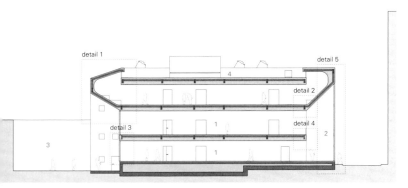

1 永久收藏品展览区 2 中庭 3 庭院 4 设备区
1. permanent collection 2. atrium 3. courtyard 4. facilities area
B-B' 剖面图 section B-B'

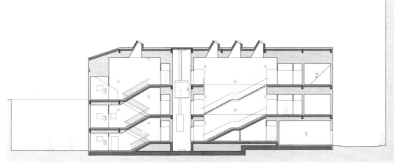

1 设备区 2 盥洗室 3 收藏品存储区&修复工作间
1. facilities area 2. lavatories 3. collection storage & restoration workshop
C-C' 剖面图 section C-C'

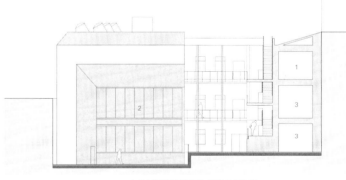

1 永久收藏品展览区 2 庭院 3 盥洗室
1. permanent collection 2. courtyard 3. lavatories
D-D' 剖面图 section D-D'

1 永久收藏品展览区 2 庭院 3 办公室 4 设备区 5 盥洗室
1. permanent collection 2. courtyard 3. offices 4. facilities area 5. lavatories
E-E' 剖面图 section E-E'

1 永久收藏品展览区 2 设备区 3 盥洗室
1. permanent collection 2. facilities area 3. lavatories
F-F' 剖面图 section F-F'

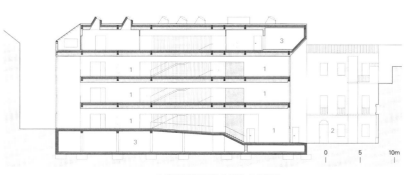

1 临时收藏品展览区 2 庭院 3 设备区
1. temporary collection 2. courtyard 3. facilities area
G-G' 剖面图 section G-G'

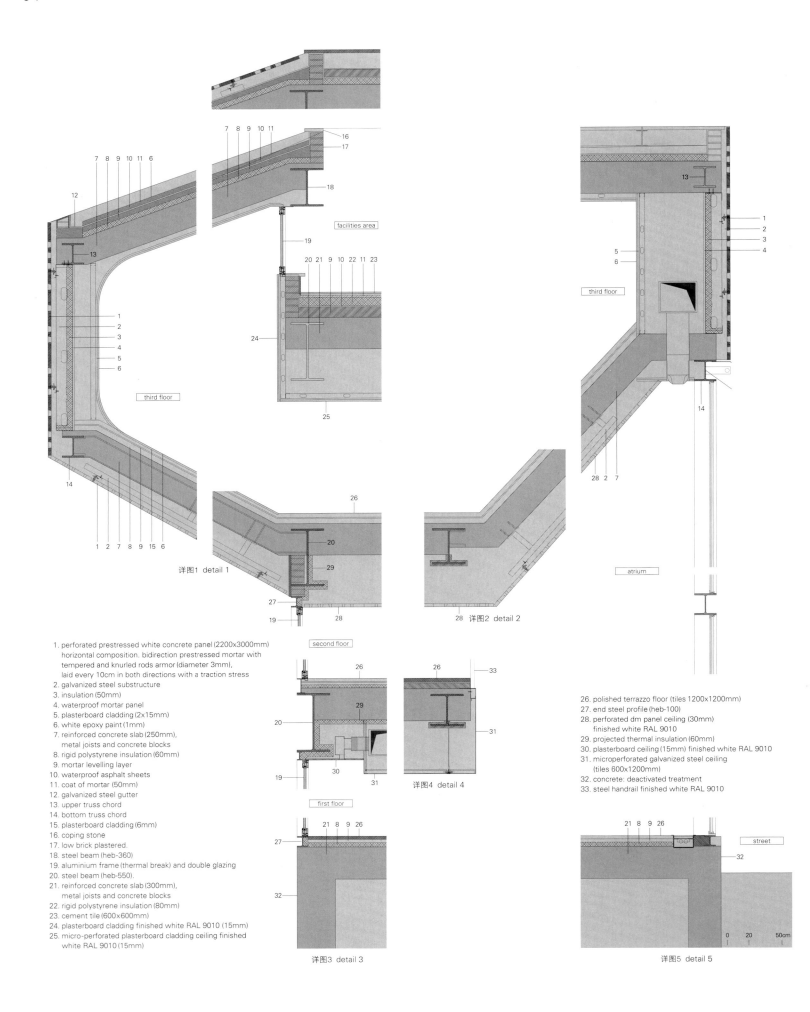

社区建筑——公共建筑与机构 In the Community—The Public and Institutional Buildings

沃特福德中世纪博物馆 Waterford City Council Architects

沃特福德中世纪博物馆是爱尔兰东南部的一个全新的建筑地标以及主要的参观目的地。

博物馆坐落在沃特福德市最古老的城区和充满活力的文化中心，即维京三角区。这个地区在过去一直被忽视，直到最近因为是城市复兴计划的核心，因此三座新的博物馆才成为全新的访客路线的核心。其中最重要的项目，一座全新的中世纪博物馆，展出了沃特福德中世纪的宝藏，且合并了中世纪的圆顶地下室。这是一个大型城市项目，建筑的正立面环绕着大教堂的背部，将两个风景优美的广场连接起来，这座建筑在内部连接着大量的临近公共建筑，也解决了建筑自身的功能需求。

一系列不同的历史和现代建筑将场地地形限制为U形，且带有一处开阔地，它横跨狭小的空隙，直接面向18世纪基督大教堂的未加装饰的东立面。在统一的表皮下，建筑包括两处主要的功能区，即中世纪的博物馆和19世纪剧院的后台设施。博物馆跨立于这两处功能区之间，且保护了中世纪时期的宗教结构，即所谓的唱诗班大厅。建筑的室内流线在维京三角区周围与外部的参观路径连接起来，手持的GPS电子多语言导航设备使内部和外部的游览没有区别。游客使用电梯在博物馆内穿梭，于顶层开始旅程，并且通过楼梯下到展示空间层的出口处。

靠近博物馆，一条玻璃通道可以提供望向下方精心设计的唱诗班大厅的视野，而悬垂的立面形成了建筑的入口。嵌入式玻璃的整个宽度都可以用作入口，带有滑动玻璃屏，使一层的室内空间朝向大教堂广场开放。

接待处作为中心枢纽，流线和室内体量围绕着它进行布局。建筑物的立面是一个长长的体块，在面向购物中心的方向为简单的矩形面，然后在水平方向上和大教堂分开，并蜿蜒于其后。博物馆连接着教堂两侧的两处精心设计的公共空间，而尾端用来突出各自的城市空间，同时利用自己的方式向周围的地区宣传着博物馆。

结构方面的设计方案是在筏式地基上建造一个现浇混凝土框架，避免简单的堆积和潜在的对周边历史建筑的损害。邓德里石材饰面由原有的中世纪教堂使用的石材制成，给人以温馨的氛围，打破了周围18世纪建筑的冷色调。一个粘土模型转化成参数化的NURBS（非均匀有理B样条）曲面计算机模型，这个模型去掉了电子表格，允许曲线随着输给切割机的加工输出信号所产生的正确切线而进行调整。一个8m高的"沃特福德妇女"雕像建在城市考古发掘中发现的一个13世纪的小型带状山脉上，且被雕刻在山墙上。

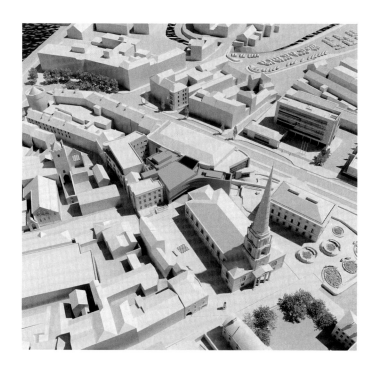

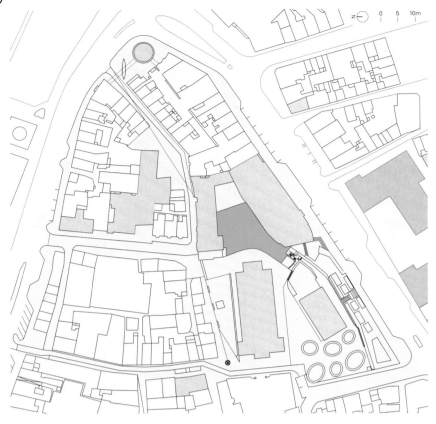

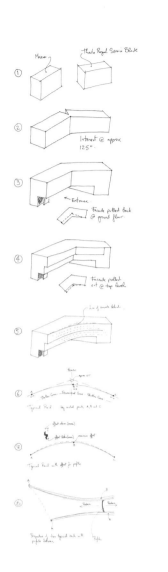
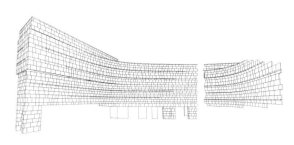
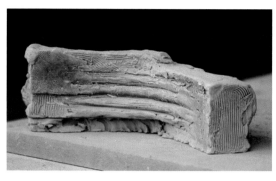

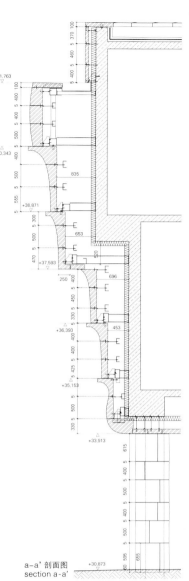

a-a' 剖面图
section a-a'

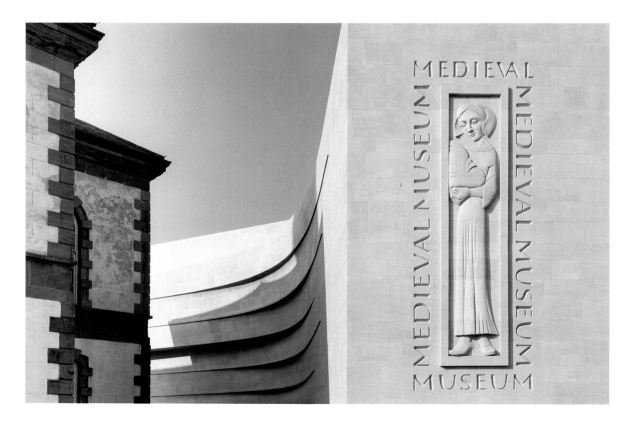

Waterford Medieval Museum

Waterford Medieval Museum is a new architectural landmark and major visitor destination in Ireland's Southeast area. The museum is located in the oldest part of Waterford City and its vibrant cultural heart, known as The Viking Triangle. This area suffered until recently from neglect and being central to its urban regeneration was the creation of three new museums forming the core of a new visitor trail. The most important project, a new Medieval Museum, displays the medieval treasures of Waterford and incorporates a medieval undercroft. This is an intensely urban project; the front facade wraps around the back of the Cathedral, creating a link between the two beautiful squares and internally the building links a number of adjacent civic buildings as well as addresses its own functional requirements.

A diverse range of historic and modern buildings constrain the site boundaries to a U-shape with the open end facing directly onto the unadorned eastern facade of 18th century Christ Church Cathedral across a narrow gap. The building envelope contains two functional components under a uniform skin: the medieval museum and backstage facilities of a 19th Century Theater. The

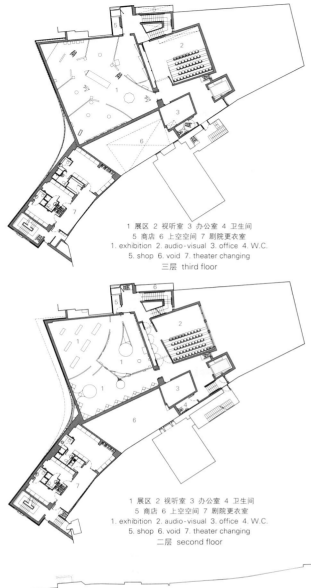

1 展区 2 视听室 3 办公室 4 卫生间
5 商店 6 上空空间 7 剧院更衣室
1. exhibition 2. audio-visual 3. office 4. W.C.
5. shop 6. void 7. theater changing
三层 third floor

1 展区 2 视听室 3 办公室 4 卫生间
5 商店 6 上空空间 7 剧院更衣室
1. exhibition 2. audio-visual 3. office 4. W.C.
5. shop 6. void 7. theater changing
二层 second floor

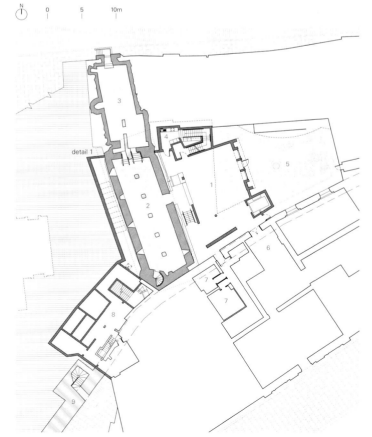

1 通用空间 2 唱诗班大厅 3 酒窖 4 厨房 5 庭院 6 市政厅长廊 7 卫生间 8 剧院后台 9 城墙
1. general purpose space 2. Choristers Hall 3. wine vault 4. kitchen 5. courtyard
6. vestibule to city hall 7. W.C. 8. theater backstage 9. city wall
地下一层 first floor below ground

1 设计中的大堂 2 门厅 3 接待处 4 商店/临时展区 5 上空空间 6 石塔
7 滑动玻璃 8 观景玻璃 9 消防路线 10 剧院后台
1. draft lobby 2. foyer 3. reception desk 4. shop/temporary exhibition 5. void
6. stone tower 7. sliding glass 8. viewing glass 9. fire route 10. theater backstage
一层 first floor

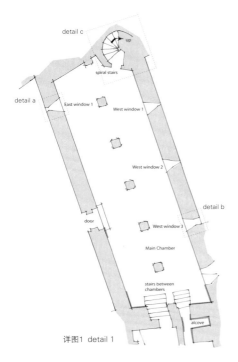

详图1 detail 1

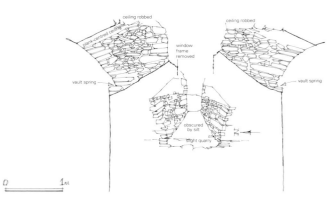

详图a detail a

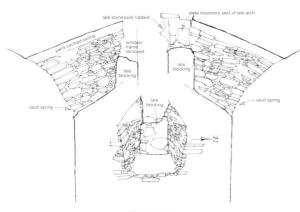

详图b detail b

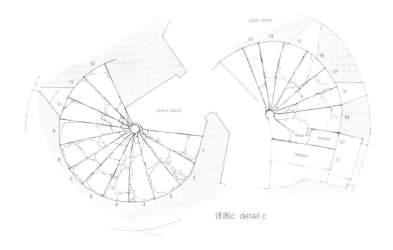

详图c detail c

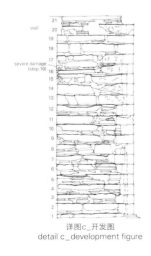

详图c_开发图
detail c_development figure

项目名称：Waterford Medieval Museum
地点：Cathedral Square, Waterford, Ireland
建筑师：Rupert Maddock, Bartosz Rojowski, Agnieszka Rojowska
项目经理：Malone O'Regan Consultants Engineers
结构、土木工程师&PSDP：Frank Fox & Associates Consultant Engineers
雕塑设计：Stephen Burke
服务工程师：WSP
消防顾问：ARUP
考古顾问：Orla Scully and Archer Heritage
立面石材承包商：S McConnell & Sons
工料测量师：Nolan Construction Consultants
制图：Waterford City Council Architects & Rojo-Studio Architects
用地面积：785m² / 总建筑面积：600m² / 有效楼层面积：1800m²
设计时间：2011 / 施工时间：2012—2013
摄影师：Courtesy of the architect - p.68, p.72, p.73, p.74, p.75
©Philip Lauterbach (courtesy of the architect) - p.66~67, p.70 bottom
©Emagine Media (courtesy of the architect) - p.69, p.70 top, middle, p.71

1 厨房　　　　　1. kitchen
2 通用空间　　　2. general purpose space
3 消防路线　　　3. fire route
4 商店/临时展区　4. shop/temporary exhibition
5 展区　　　　　5. exhibition
6 上空空间　　　6. void

museum straddles and protects a conserved ecclesiastical medieval structure known as the Choristers Hall. The internal movement through the building links with the external visitor trail around the Viking Triangle and the hand-held GPS electronic multilingual guides make no distinction between inside and outside. Visitors move through the museum by using the lift to start their tour on the top floor and use the stairs to drop down to the next level on exiting the exhibition spaces.

Approaching the Museum, a glass pavement provides views to the Choristers Hall below and the facade overhangs framing a gateway or portal into the building. The entire width of recessed glazing can be used as entry, with sliding glazed screens being used to open the ground floor to Cathedral Square.

The reception desk acts as a central pivot object around which the circulation and internal volumes are organized. The facade is a long block which exposes a simple rectangular face to the Mall and then sweeps around the back of the Cathedral splitting horizontally. It connects the two fine public spaces on either side of the Cathedral and ends are used to reinforce their respective urban spaces as well as advertise the Museum to its approaches.

The structural solution was to form an in-situ concrete frame placed on raft foundations avoiding piling and potentially damaging adjacent historical buildings. The warm Dundry stone facing follows from its use in the original medieval Cathedral and provides a break from the cool style of 18th century surrounding structures. A clay working model was converted to a parametric NURBS surface computer model that was driven off a spreadsheet allowing the curves to be adjusted for correct tangents with the processed output fed to cutting machines. An 8m high figure, "the Waterford lady", based on a tiny 13th century belt mount found in city excavations was sculpted on one gable.

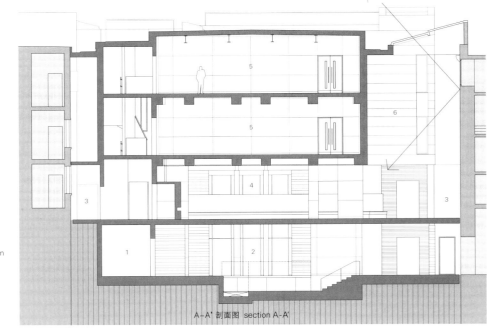

A-A' 剖面图 section A-A'

手工造纸博物馆

TAO

博物馆建造在云南高黎贡山下新庄村边的优美风景中，这里是中国西南部一处世界生态保护区。这个村庄拥有悠久的手工造纸传统。为了展出手工造纸的历史和文化，博物馆将包括长廊、书店、工作空间以及客房。博物馆被认为是一个微型村庄，由几座小型建筑构成。其空间理念是当游客在博物馆内穿梭时，内部长廊和外部优美的田园景观之间能够不断转换体验，以此来提示造纸和环境密不可分。

设计旨在建造一座对气候和环境做出友好反应的建筑。本地的材料，如木材、竹子和手工制作的纸和火山石分别应用在外饰面、屋顶、室内饰面和地面中。随着时间的流逝，这些材料将会褪为一种与景观更加相融的颜色。整个施工过程将本地材料、技术和手工艺的使用最大化。建筑结合了传统的木结构体系，采用无钉的榫卯来进行连接，细节设计十分现代化。这座建筑全部是由当地工匠完成建造的。

Museum of Handcraft Paper

The museum is situated in a beautiful landscape next to Xinzhuang Village under Gaoligong Mountain of Yunnan, a world ecological preserve area in the southwest of China. The village has a long tradition on handcraft paper making. To exhibit the history and culture of paper making, this museum will include gallery, bookstore, work space and guest rooms. The museum is conceived as a micro-village, a cluster of several small buildings. The spatial concept is to create a visiting experience alternating between interior

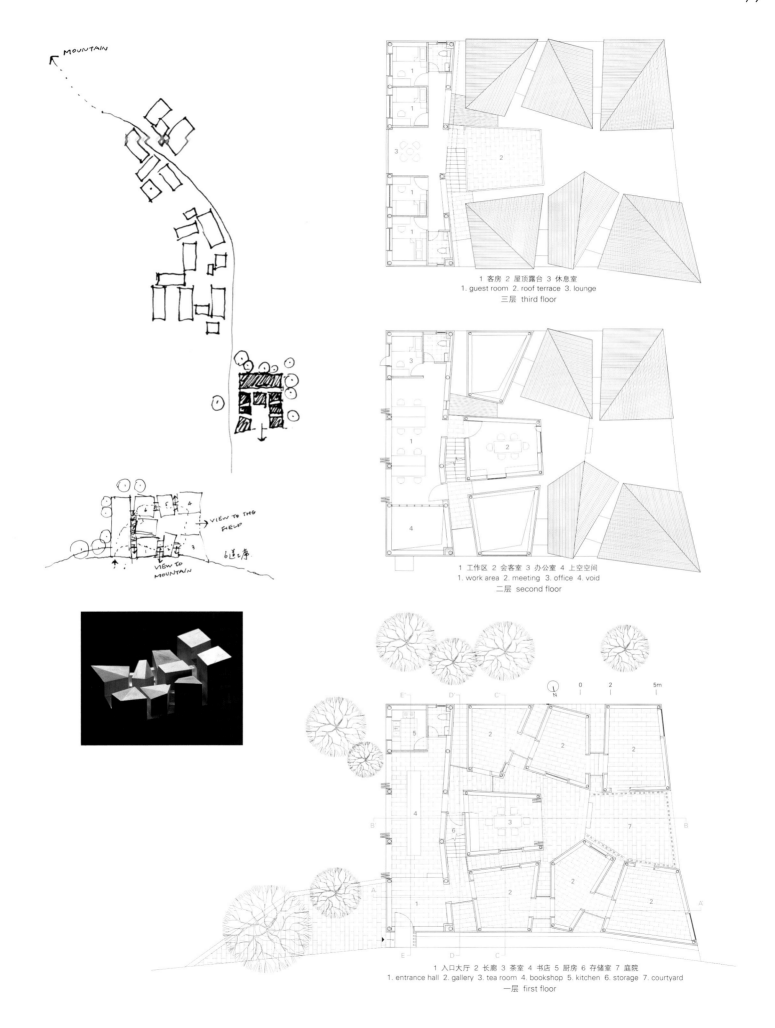

of galleries and landscape outside when a visitor walks through the museum, so as to provoke an awareness of the inseparable relationship between paper making and environment.

The design is aimed at making a climate responsive and environment friendly building. Local materials such as wood, bamboo, handcraft paper and volcano stone are used for exterior finish, roof, interior finish and floor respectively. As time passes, these materials will worn and fade into a more harmonious color with the landscape. The construction is to maximize the usage of local materials, technique and craftsmanship. The building combines traditional timber structural system featuring nailless tenon(Sunmao) connection and modern detailing. It was built completely by local builders.

1. 20mm volcano stone/30mm cement motar layer/rammed earth
2. 50mm volcano stone/200mm pebble concrete/50mm volcano stone
3. 40x300mm volcano stone(various length)/ with wholes for ventilation
4. 20mm fir wood board/20x20mm horizontal wood strips/fixed on 20x20mm vertical wood strips/5mm waterproof membrane/20mm plywood 50x50mm horizontal wood strips fixed on / 70x70mm vertical wood strips/90mm cavity/30x50mm wood frame with 450x450mm spacing/handcraft paper glued on the frame
5. niche for exhibition
6. 5+5mm safety glass
7. wood beam
8. 20mm wood board/2mm galvanized iron sheet(waterproof)/20x20mm wood strips/ fixed on 50x60mm wood strips which are parallel with the water/drainage direction/ bamboo
9. gutter

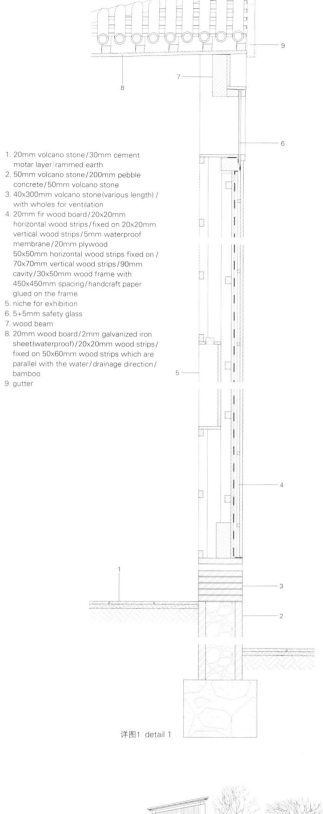

详图1 detail 1

西立面 west elevation

东立面 east elevation

北立面 north elevation

南立面 south elevation

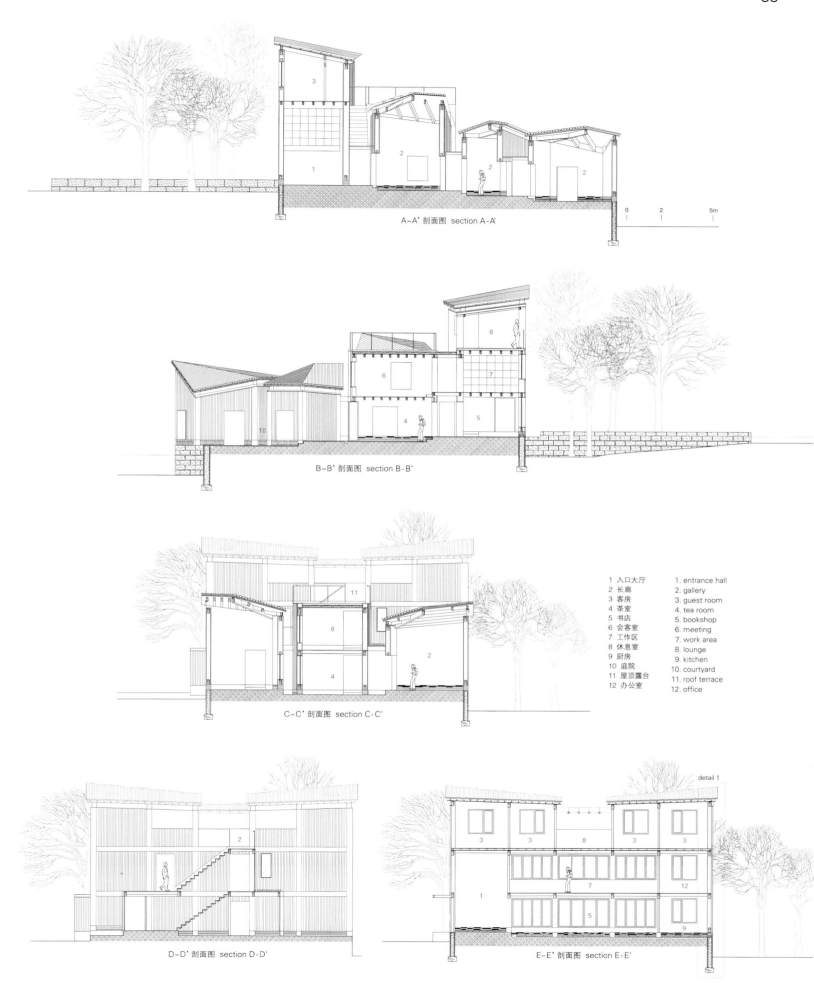

项目名称：Museum of Handcraft Paper
地点：Xinzhuang Village, Tengchong, Yunnan, China
建筑师：Hua Li
项目团队：Hua Li, Huang Tianju, Li Guofa, Jiang Nan, Sun Yuanxia, Xu Yinjun, Yang Hefeng
施工组：Local farmer builders led by Long Zhanwen
甲方：Committee of Gaoligong Museum of Handcraft Paper
用途：museum, work space, guest room
用地面积：315m² / 总建筑面积：193m² / 有效楼层面积：361m²
结构：timber construction with traditional Chinese tenon connection / 屋顶：bamboo
立面材料：timber board, volcano stone
室内饰面材料：handcraft paper / 地面材料：volcano stone
设计时间：2008.4—2009.5 / 施工时间：2009.5—2010.12
摄影师：courtesy of the architect - p.76, p.77, p.87
©Shu He(courtesy of the architect) - p.78, p.80, p.82, p.84, p.85, p.86, p.88~89

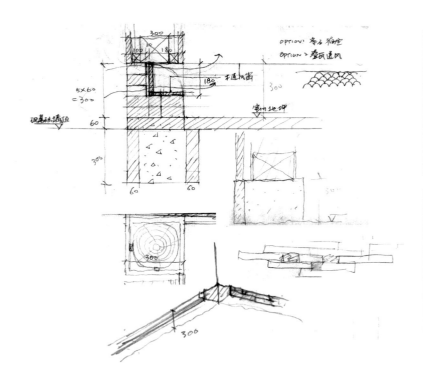

澳大利亚恐龙时代博物馆

Cox Rayner Architects

澳大利亚恐龙时代博物馆是一位牧羊人David Elliott脑海中的理念，他于十年前在自己的住处发现了一些奇特的岩石，后来被证实是全世界最重要的大型恐龙化石，而他和全家人从此便致力于研究古藻类学。数年来，挖掘、收集和保护化石的过程吸引了无数的志愿者、小镇居民以及游客的参与。而建造这座小型博物馆的目的就是使游客能够观赏场地内的发现，以推动当地经济，并且为正在进行的保护工作建造一些有利设施。

场地是一处较为偏僻的台地，约75m高，可以俯瞰原有的和将来可能发现恐龙化石的冲击平原。这处台地是由一位当地的农民捐献的，已不再用于放牧。由于资金的缺失，以及对Elliott工作价值的信任，建筑师同意无偿地设计这座博物馆。而建造商woollams公司，也将一半的精力投入到帮助Elliott家庭以及当地的支持者建造博物馆中来。

这座建筑包含一个中央解说大厅，大厅面向一个露台开放，露台延伸至悬崖的边缘。此外，博物馆还包括一处接待区、行政区、咖啡室、便利设施以及一间"特形房"。最大的空间便是化石展出的地方，里面的气候可控制，并被赋予了解说。

该项目的设计理念来自于台地的景观，尤其是深层岩石缝隙的特征。建筑被构思为一组巨石结构，人们通过一个狭窄的孔径进入其中。在里面，人们可以通过其他的孔径瞥向天空、高原和远处的平原。孔径将这种包容感置于断裂的地质结构中。

外部，建筑如同一条变色龙，对面的悬崖延伸至其上方。嵌入地面的、潮湿的现浇倾斜混凝土墙体与场地的朱红色泥土，使其在材料方面实现这一效果。施工过程使墙体具有胶乳纹理，这是一种来自于观察胶乳模具（用来放置发现的恐龙化石）在切割时是怎样保留化石痕迹的方法。

抽象的巨石特征由带有棱角的且相互连接的墙体进行突出，每面墙都通过生锈的铁楔与相邻的嵌板相连接。生锈的铁楔也广泛地应用在

项目名称：The Australian Age of Dinosaurs Museum
地点：Winton, Queensland, Australia
建筑师：Cox Rayner Architects
项目团队：Michael Rayner, Casey Vallance, Justin Bennett
合作者：Bligh Tanner, Cushway Blackford & Associates, Consulting Coordination, Chris Battersby (Woollam Constructions) Museum
甲方：Australian Age of Dinosaurs
有效楼层面积：394m²
项目开始时间：2008.2 / 竣工时间：2012.4
摄影师：©Christopher Frederick Jones (except as noted)

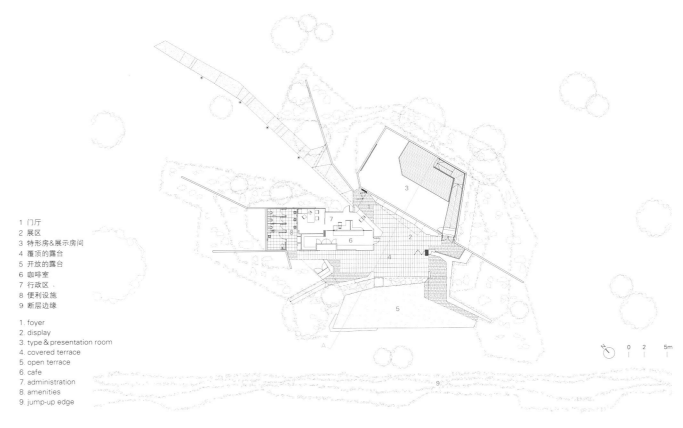

1 门厅
2 展区
3 特形房&展示房间
4 覆顶的露台
5 开放的露台
6 咖啡室
7 行政区
8 便利设施
9 断层边缘

1. foyer
2. display
3. type & presentation room
4. covered terrace
5. open terrace
6. cafe
7. administration
8. amenities
9. jump-up edge

滑动门板、容纳发电机的塔结构以及人工穿孔的嵌板（使用等离子切割机来切割从地面投射的树影子中汲取的图案）中。当它们组合在一起时，墙体和围屏与场地的色调相似，共同体现景观的永久性和脆弱性之间的对比。

除了胶乳铸型法，出于各种用途，其他几种可见的本地技术也应用其中，如利用马具皮革来包裹门柄。

该项目是建筑师认为最有意义的设计之一，即一座为完全奉献给其追求的甲方而设计的建筑，且大部分的施工是由甲方完成的，他们之前完全没有经验。文化旅游使这个项目获得了成功，刺激了沿着悬崖边缘建造一座更有恢宏抱负的博物馆的可能性。

Australian Age of Dinosaurs Museum

The Australian Age of Dinosaurs Museum is the brainchild of a sheep grazier, David Elliott, who discovered strange outcrops on his property over a decade ago. They turned out to be some of the world's most significant finds of large dinosaur fossils, and he and his family have devoted their working life to the art of palaeontology ever since. The process of digging, collecting and conserving the fossils has involved countless volunteers, townspeople and visitors over the years. The idea of creating a small museum was to enable visitors to appreciate the discoveries in situ, to boost the local economy and to create quality facilities for ongoing conservation work.

The site chosen was a remote mesa some 75 meters high overlooking the alluvial plains where existing and future dinosaur finds lie. The mesa was donated to the cause by a local farmer, it being unusable for grazing. Due to absence of funds, and the belief in the value of the Elliott's work, the architects agreed to do the project pro bono. The builder, Woollams, partly donated its services to help the Elliott family and their local supporters to construct the museum.

The building comprises a central interpretive space that opens out to a terrace extending to the cliff edge, a reception and ad-

挖掘沟渠 carving trenches

为混凝土上色,使其与地面颜色一致
coloring concrete with earth

印上胶乳纹理 imprinting latex textures

完工的预制墙板
completed precast walls

安装第一块板 erecting the first panel

将互相连接的墙板支撑起来 propping interlocking walls

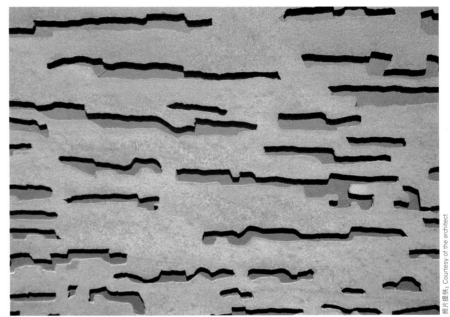

照片提供: Courtesy of the architect

墙板铺陈在地面上,等待安装 wall panels laid out for assembly

用白垩粉涂抹图案,以突出铁板上的穿孔
chalking patterns for perforations on iron sheets

ministration area, cafe and amenities, and a "type room". This largest volume is where the fossil finds are displayed in climate-controlled conditions, and interpretive talks are given.

The design concept drew inspiration from the mesa landscape, especially its character of deep rock fissures. The building was conceived to be like a cluster of boulders, entered via a narrow aperture. Within, other apertures afford glimpses to the sky, the plateau and the plains beyond, imparting the sense of containment inside a fractured geological formation.

Externally, the building was intended to also appear like a chameleon and an extension of the cliff faces it sits above. This aim was materially achieved by on site poured tilt-up concrete walls embedded while wet with the vermillion red earth of the site. The process involved texturing the walls with latex fabric, a method that was derived from observing how latex moulds, used to cradle the found dinosaur bones, retain the fossil imprints when cut away.

The character of abstracted boulders is accentuated by the walls' angular, interlocking shapes, each connected to neighboring panels by rusted iron wedges. Rusted iron is also extensively used for sliding door panels and a tower enclosing the generator, the hand-perforated panels using plasma cutters to cut patterns taken from tree shadows across the ground. Together, the walls and screens approximate the range of hues of the site, and they were conceived to embody the contrasting permanence and fragility of the landscape.

In addition to the latex molding, several other observed local techniques were adapted to various purposes, such as wrapping door grips in leather using saddlery.

The project is one of the most rewarding projects and the architects have been engaged in designing a building for a clients utterly dedicated to their own pursuit, and largely constructed by them with no previous experience. The cultural tourism success of the project has stimulated the possibility of creating a more ambitious museum along the cliff face edge.

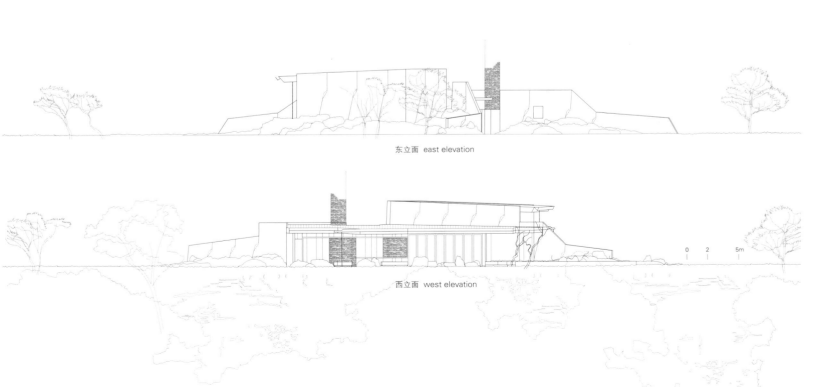

东立面 east elevation

西立面 west elevation

照片提供：Courtesy of the architect

A-A' 剖面图 section A-A'

个性化建筑
The Individuality and

医疗和教育类的建筑物有很多相似之处,尽管它们最初可能看起来大相径庭。作为向当地居民提供重要服务的公共建筑物和公共机构,它们必须努力克服一些相似的问题。公共机构通常被规则、官僚主义以及个性特点缺失所限制,那么在这种情况下,应如何将这些建筑物推回到原有状态呢?

下面介绍的这些建筑物就展示了一系列的这类模式,建筑师和他们的甲方正是通过这些模式来寻求方法,以削弱典型的医疗及教育机构建筑的公共机构特点,并且采用一系列的关键措施来解决这一问题。

Health and education buildings have much in common, even as they might initially appear very different. As public buildings providing important services to their local neighbourhoods and as institutional buildings, they have to grapple with many similar issues. Where institutions are often defined by their rules, bureaucracy and the lack of personalisation that entails, how does the architecture push back?

This collection of buildings reveals a series of patterns in the ways that the architects and their clients have sought to mitigate the more institutional aspects of typical health and educational buildings, employing a number of key strategies in order to do so.

Råå日托中心_Råå Day Care Center/Dorte Mandrup Arkitekter
农场幼儿园_Farming Kindergarten/Vo Trong Nghia Architects
Pies Descalzos学校_Pies Descalzos School/Mazzanti Arquitectos
Zugliano的某所学校_School Complex in Zugliano/5+1AA
El Guadual儿童早期发展中心_El Guadual Early Childhood Development Center/Daniel Feldman & Iván Dario Quiñones
木质牙科诊所_Timber Dentistry/Kohki Hiranuma Architect & Associates
哥本哈根癌症防治与康复中心_Copenhagen Center for Cancer and Health/Nord Architects
利迈健康中心_Medical Care Center in Limay/Atelier Zündel Cristea
机构模式_Institutional Patterns/Alison Killing

机构模式

医疗中心、学校和幼儿园等都是重要的公共建筑,在为当地社区提供重要服务的同时也潜在地成为社区的重要中心。除此之外,虽然,这些规划从表面上看似乎有本质上的差别,且内部容纳的设施也有所不同。但是,近距离观察,它们必须努力克服一些相同的问题以满足那些需求,并且必须采用相似的策略来处理这些问题。

在少数的重大事情方面,这些建筑还是有共同点的。第一,教育和医疗建筑在更大程度上以公共机构的特点为中心,即包含所有的规则、官僚主义,并且缺乏个性特点。而这些特点又都不能被看作是积极的特性,在这种情况下,应如何将这些建筑物推回到原有状态呢?其次,隐私和安全问题又要求这些建筑物的布局向内集中。尽管如此,但是必须再次申明,这一建筑趋势大多呈现在健康护理类建筑中。那么,如何让这些建筑物更好地与外界相接触呢?

由平沼孝启建筑师事务所设计,位于日本大阪箕面的木质牙科诊所

Institutional Patterns

Health centers, schools and kindergartens are important public buildings, providing key services to their local communities and potentially serving as important centers for those communities. Beyond that, however, their programs would appear to be quite fundamentally different, demanding very different things of the buildings that house them. On closer inspection, they have to grapple with some of the same problems in aiming at meeting those demands and employ similar strategies to deal with them. They have a few, quite significant things in common. The first is that education, and health buildings to an even greater extent, center around an institution with all of the rules, bureaucracy and the lack of personalization that entails. None of these are seen as particularly positive characteristics, in which case, how does the architecture push back? And then there are the privacy and safety issues, which tend to produce quite inwardly focused buildings, although again, that tendency is most pronounced in the health care buildings. How then do these buildings engage with the outside world?

The Timber Dentistry in Minoo, Osaka Prefecture in Japan by Kohki Hiranuma Architect & Associates is perhaps the most open building showing here. The clinic is located in a quiet residential suburb and takes several of its cues regarding massing and form

the Institution

可能是这里最开放的建筑物了。这个诊所坐落在一处安静的住宅区的边缘处,并把附近住宅的体块和形式作为设计依据。它们的规模大致相同,屋顶使用了相邻建筑的三角形屋顶形状,并且逐渐向下扭曲,在建筑物的另一端呈直线状。所产生的动感性使建筑与周围建筑大不相同。

该诊所的不同点十分引人注目,亮白色的诊所与周围住宅柔和的灰色和棕色房屋相映衬,甚至其立面上的窗户和开口也有所不同。从各个方面来说,这种对比是非常合适的,因为它有助于标志出诊所的不同功能以及有用的社区服务。诊所内明亮的一层大厅看上去很开放,敞开怀抱欢迎访客,但周围住宅的前面设有围墙,立面看上去更加封闭,并且窗户紧闭,它们似乎要将整个世界控制在可触及的范围之内。等候室直接与道路相对,访客就在这处毫无隐蔽的地方等候预约。医疗机构通常都比这个诊所更在意拜访者的隐私问题,即便是等候室也是如此。尽管比起癌症治疗中心,隐私问题对于牙科的一些常规治疗来说并不是那么迫切的需求。然而,建筑的外观可能具有欺骗性,这座建筑物实质上比其展现的封闭的多。比如,这一建筑的较高楼层就是对街道完全封闭的,而从内部的房间(主任办公室和员工的房间)都可望向位于建筑物后面的玻璃露台,并从这里采光。与此相似,诊所一层的大部分房间也都位于等候室的木质后墙的后面,隐藏起来。而灭菌室、X光室和治疗室也位于这些隐蔽的地方。这就像是一种戏法,从街面上看,最初看起来最开放的地方其实也是最隐秘的地方。

由诺德建筑师事务所设计,位于哥本哈根的癌症防治与康复中心也经历了相似的困境。作为社区内的一个重要设施,它是一座高度可见的建筑,并且成为一个标志,以此引起人们对癌症的认识,同时避免人们对癌症病人的歧视。最终,建筑师们希望通过一些措施来消除该中心给人们的公共机构的感觉。许多医疗护理机构特别是大型的医院给人的感觉就是在这些地方癌症病人必须接受大量的医疗护理、放射性治疗以及化疗等。

人们来这到一中心接受的是建筑师声称具有"温和"元素的癌症治

from the adjacent houses. It is approximately the same size, and the roof borrows the shape of the gable of the house next door, gradually twisting down to become more rectilinear at the building's opposite end. It's a move from a building that is otherwise quite different from its neighbors.

The differences are quite striking, from the bright white of the clinic, set against the muted gray and brown of the houses, to the contrasting approach to windows and openings on the facade. That is in many ways appropriate – it's helpful to signal the presence of a different function and useful local service. The glazed first floor of the clinic appears open and welcoming to visitors, while the houses, with the fences in front, more closed facades and shuttered windows seem to determine to hold the world at arm's length. The waiting room faces directly onto the street, with visitors waiting in this fish bowl for their appointment. Health buildings are usually much more attentive to visitors' privacy than this, even for the waiting rooms, although this is probably a much less urgent consideration with the more routine work of a dental surgery, than say, that of a cancer treatment center. Appearances can be deceptive however, and this building is much more closed than it appears. The upper floor, for example, is completely closed to the street and the rooms that are located there – the director's office and the staff room – look out onto and are lit from a glazed terrace at the back of the building. Similarly, the majority of rooms on the first floor are tucked away out of sight behind the wooden back wall of the waiting room. It's here where the sterilization room, x-ray and treatment rooms are located. It's almost a sleight of hand – what initially seems to be the most open building on the street is also its most intensely private.

Nord Architects' Center for Cancer and Health has to walk a similar tightrope. It is highly visible as an important facility for the community; the building was also conceived of as an iconic piece of architecture, with the idea of raising awareness of cancer. On the other hand, there was a need to avoid stigmatizing patients. Finally, the architects also wanted to avoid the institutional feeling common to many health care buildings and certainly to the sort of large hospitals where the center's patients would have to go for much of their medical care, their radiation therapy and chemotherapy.

This center then is where people come for what the architects refer to as the "softer" elements of their cancer treatment. It offers physiotherapy, dance and fitness classes, cooking lessons, space for support and discussion groups. It's also non-residential, with patients and their families dropping in to access services as they need them. All of these things make it easier to depart from the more institutional, technocratic and bureaucratic factors which

木质牙科诊所，大阪，日本
Timber Dentistry, Osaka, Japan
照片提供：©Kohki Hiranuma (Satoshi Shigeta)

疗。该中心为患者提供物理疗法、舞蹈和健身课程、烹饪课程，还有援助和小组讨论室等。同时它又是一座非住宅性建筑，病人及他们的家人可以随时走访，以接受他们所需要的服务。所有的这些特点使它更容易与那些典型的大型医疗护理类建筑（如医院）区分开来，摆脱了更多的制度、技术以及官僚主义的束缚。

我们有充分的理由说明为什么医院应该改变其自身的工作方式。传统医院的设计通常需要满足构建大型公共机构的布局要求，以及现代医疗设施的技术要求。而现代医院应主要以服务、设备和专业技术为中心，以期变得更高效，并能为更多的人提供周到服务。传统医院里无尽头的走廊和人们必须四处找路等问题都是由建筑及其布局规模十分庞大造成的。布局的规模以及划分为不同部门的方式等因素都可能包含在任意一个病人的治疗过程中，而这也将突出其缺乏人性化且具有公共机构特性的感觉。

诺德建筑师事务所的哥本哈根癌症防治与康复中心的规模较小，并且布局相对狭窄集中，因此在很大程度上避免了上述问题。这种设计比传统的医院更加自由，因为它并非一定要拥有大型的、灵敏度极高的医疗设备。即使不是很容易，那么形成一种建筑师认为的具有更加温馨且友好的感觉，是十分必要的，这不仅仅要拒绝冷酷、没人情味的医院风格，还要为人们带来更快恢复健康的、积极的期望。

这座癌症中心并不是特别典型的小建筑，尽管在常规医疗护理环境中，它算得上是小建筑。在规模和形态方面，它是一座非常具有标志性建筑，但是也要灵活地处理外向型设计的意愿和采用智能的方式来建造更小规模且更安静的建筑的需求之间的张力。建筑物的形态实质上是一个封闭的庭院，被分解成若干体块，从屋顶轮廓线方向看去，如同一系列带有斜屋顶的房子。而这个受日本折纸艺术启发的屋顶设计和整齐划一的粉刷白墙又将一座座分散的房屋融合成一座整体建筑。该建筑同时兼顾大规模和小规模建筑的理念，并使它们能够完美结合。而一旦进入建筑物内部，我们还会发现相似的大规模和小规模设计。中心内部有一组志愿者，专门接待到访的病人并为他们指路，这一举动无疑使病人的体验私人化。但是，建筑物本身也突出了这一点。它围绕着一个不规则的中心庭院而建，这一位置创造了一些稍微偏离了主要区域的、更私密的空间。这样访客就可以在一个公共机构的内部找到一处私人空

typically shape larger health care buildings, like hospitals. There's a good reason why hospitals end up the way that they do. Their design is frequently driven by the organizational demands of a large institution and by the technology necessary to a state of the art medical facility. The modern hospital is predicated on centralizing services, equipment and expertise in order to be more efficient and to be able to prove those services more affordably to a greater number of people. The sheer size, of both organization and building, results what leads to those endless corridors and problems with finding your way around. The scale of the organization and the way in that it then needs to be broken down into departments, which may be involved in any one patient's treatment, increase the sense of depersonalization and the institutional feel. The smaller scale of Nord's center and its narrow focus mean that to a large extent it doesn't have to contend with these issues. The design can also be somewhat freer than the conventional hospital since it doesn't have to accommodate sensitive, and hefty medical equipment. If not easy, to create the more homely, welcoming feel that the architects felt was needed, not only as a rejection of the cold, impersonal hospital, but also with the positive intention of helping people to get better faster.

The cancer center isn't a particularly small building, although in the context of health care it might reasonably be considered small. In both its size and its form it is pretty iconic, but also deals with the tension between its desire to be more extroverted and the need for a smaller scale, quieter building in a clever way. The form of the building, essentially a closed courtyard block, is broken down into what would appear from the roofline to be a series of pitched-roof houses. The origami inspired roof and unifying white stucco reform the block as one building and it contrives to be both small scale and large scale at the same time, a neat move. Once inside, there's a similar play of small versus large scale. There's a team of volunteers at the ready to welcome patients as they arrive and help them find their way and this of course personalizes the experience. But the building also does its part. It is built around a central, irregularly shaped courtyard. This move creates a number of more intimate spaces at a slight remove from the main areas, in which visitors might find peace, quiet and privacy – a space for the self in the midst of an institution.

Atelier Zundel and Cristea's Medical Care Center in Limay, France is a much bigger facility. It's a residential care center for people with a range of disabilities, from mental and sensory handicaps, to those with motor disabilities. The building is organized around four residential units and with its single story, it extends to cover a

利迈健康中心，法国
Medical Care Center in Limay, France

间，这里平和、安静并且私密。

由Zundel & Cristea工作室设计的法国利迈健康中心是一个更大的医疗设施。它是一个住宅性保健中心，主要为一些在精神、知觉或是运动方面有障碍的残疾人服务。这一建筑围绕着四个居住单元而布局，且只有一个楼层，建筑向外扩展，覆盖了一处相当大的表面区域。正如建筑师们所说，为这么大规模的建筑提供服务的走廊（特别是那些贯穿同一楼层而不是与高层连接的走廊），不再成为该建筑物的主要元素。在这种情况下，建筑师们只能尽力使这些走廊成为积极的元素。

该建筑由六个小庭院隔开，这些庭院不仅成为居民观赏的花园，同时对建筑物采光和通风也是非常必要的。其内部的走廊网络沿着室外空间的边缘延伸，并将很多密闭的房间推向建筑的边缘处。这样走廊也就变成了明亮且通风的场所，而通常情况下，走廊都位于建筑物内部，并将立面空间留给走廊通向的房屋。在这里，走廊形成了属于自己的空间，在变窄并继续向前延伸后，又在关键地方拓宽。通过这种方式创造的多样化环境为会议、集会和闲逛创造了舒适的空间，而在传统的医院中，走廊通常显得很不人性化，甚至有些无情。

瑞典赫尔辛堡Rää日托中心可以说是医疗护理中心的一个缩影。为了使建筑物更具有特点，建筑物周边的规模也进行了规划，这对于迎合它的少数居民来说是非常重要的。该建筑为一层结构，呈T形，灵感来源于其周围的沙丘景观。其"T"形结构的两个短小的分支与附近的小学连接起来，这样就在两所学校之间形成了两个可以免受海风洗礼的操场。

幼儿园的孩子们被分成四组，每一组在中心内部都有专用空间，即小组房间，立面看上去似渔民小屋的山墙。这些小组房间嵌在建筑的更大空间中，并且通过半透明的"书架"式墙壁将自身与周围的公共空间隔开。每个小组专用的流线空间完全不见，取而代之的是充满缤纷色彩和多样活动的流动空间。

第二座幼儿园是越南边和市的农场幼儿园，由Vo Trong Nghia建筑师事务所设计，拥有非常有趣的流线。流线遵循着向下循环的屋顶的路线，形成一个连续的带状结构。教室建在建筑的外围，使一个开放的凉廊围绕着庭院的内部，作为流线空间，并为孩子们提供与朋友闲逛和玩耍的场所。孩子们也可以进入屋顶的菜园，这里是专门用来教授他们有

fairly significant surface area. As the architects themselves say, the corridors that you need to service such a large building (especially when they spread across one floor, rather than being stacked upwards) end up becoming a major element of the building. Here, they have worked hard to make these corridors a positive feature. The building is punctuated by six small courtyards, which offer gardens for the residents, but are also necessary for lighting and ventilating the building. The network of corridors skirts the edges of these outdoor spaces, with many of the closed rooms pushed to the outer perimeter of the building. These corridors then become light, airy spaces, where often the temptation is to push them to the interior to leave the facades free for the rooms the corridors give access to. Here they get to become spaces in their own right, opening out at key points before narrowing and continuing on their way. The variety of environments that they have created in this way make comfortable spaces for meeting, gathering and hanging around, out of what could have been impersonal and slightly relentless, as corridors through hospitals often are.

The Råå Day Care Center is almost a miniature version of the medical care center, and has size on its side when it comes to personalize the building, something especially important to the small inhabitants it caters to. The building takes inspiration for its form from the surrounding dune landscape. It is single story and T-shaped in plan, the short stem of the "T" joining on to the adjacent junior school. This move creates two playgrounds, between the two schools and sheltered from the wind coming off the sea.

The kindergarten has four groups of children and each has their own dedicated space within the center – a group room which manifests the facade as the gable of a fisherman's cottage. These group rooms are inserted into the larger space of the building and shut themselves off from the communal space around by means of semi-transparent "bookcase" walls. Dedicated circulation space is completely done away, in favor of a fluid space full of riotous colors and activities.

A second kindergarten, the Farming Kindergarten by Vo Trong Nghia Architects in Bien Hoa, Vietnam, has a lot of fun with its circulation, as it follows a roof that loops down and round on itself, forming a continuous strip. The classrooms are pushed to the outer perimeter, leaving an open loggia to run around the inside of the courtyards, serving as the circulation space, but also as somewhere to hang around with friends and play. It's also possible to go up on the roof, where there is a vegetable garden, expressly for the purpose of teaching the children about food and growing. Kindergartens often work hard to be welcoming and personal,

Råå日托中心，赫尔辛堡，瑞典
Råå Day Care Center, Helsingborg, Sweden

关食物和生长的知识的地方。幼儿园通常在呈现出内向型布局的同时，也尽力展现友好且更具个性的氛围。而该建筑正是这样的一个奇妙的、有趣的典范。

随着医疗护理机构和教育机构规模的变大，其所面临的挑战也更大。我们也开始能够从为更大孩子准备的更大型学校建筑中看到公共建筑的变化轨迹。更多的房间、更多的走廊便需要潜在的更多的公共机构体验。下面要介绍的两座建筑物也具有明显的内向型外观，即包围学校室外游乐区的、由教室组成的一个（或者若干个）环形结构。

由5+1AA事务所设计的Zugliano的某所学校几乎占据了它的整个校址，其建筑四周留出的空余地的长度仅不到一米。学校屋顶形态的灵感来源于其附近的山脉，而建筑团队的理念之一是将景观投射在建筑的立面上。外墙粉刷成暗灰色，这样当教室的门向这一侧打开时，就会给人一种整体的封闭感，有时甚至是一种不友好的感觉。

学校的内部世界却截然相反——活泼开放，并且丰富多彩。内向型布局对需要保护自己的内部事务的学校来说并非不合适。学校中心有一个庭院，无论是表面还是实质意义方面，它都是学校的中心，因为它不仅是一个操场，也是学校集会的地方。连接教室的走廊沿着庭院的边缘延伸，但是时而变宽，时而变窄，创造出一处变化的、不规则的空间，而这比更直且更窄的标准流线更有助于人们逗留和集会。构成山体景观式屋顶的不规则板材，从意想不到的角度伸出，位于操场的上方，并在其周边形成各种各样的空间。而丰富多彩的内部空间又将这一公共机构分解为一些小的单元，使每个人都可以更容易地找到属于自己的空间。

由Giancarlo Mazzanti设计的哥伦比亚卡塔赫纳市Pies Descalzos学校，比前面所提到的学校的规模更大，是其附近街区的一座主要建筑物。建筑内有三所学校——幼儿园、初中和高中，以及大型学校所必需的辅助性的图书馆、礼堂以及运动设施等。每一所学校都围绕着各自的六角形庭院布局，庭院是一处荫蔽的空间，里面种满了植物以创造凉快的微气候。虽然这个公共机构的庞大规模及其重复的形态可能使其缺乏个性，但是它也拥有一个非常娱乐性的特色。例如，庭院周围有一个蜿蜒向上的坡道，为孩子们提供了理想的游戏和奔跑场所。科学实验室采用悬臂式从主建筑伸出，大型窗户将美景纳入其中，并且把后方区域和更广阔的景观连接起来。

even as they tend to be fairly inward looking. This is an incredibly playful example.

As the health care buildings, and education buildings get larger they become more challenging. We start to be able to see the traces of the institution in the larger school buildings for older children. With more rooms and more corridors to deal with a potentially more institutional experience is needed. These following two buildings are also intensely inward looking, a ring, or rings of classrooms encircling the schools' outdoor play areas.

The School Complex in Zugliano by 5+1AA occupies its entire site, the building stopping barely a meter from its perimeter on any one side. The form of the roof was inspired by the nearby mountains, one of the team's concept sketches showing this landscape projected on the building's elevation. The exterior walls are painted a dark grey and while classroom doors open out onto this side, the overall effect is very closed, a little unwelcoming even.

The inner world of the school is the polar opposite – lively, colorful and open. This introversion isn't inappropriate for a school which needs to be protective of its charges. The courtyard is at the center for the school, both literally and metaphorically – it isn't just a playground, but it is also a place where the school gathers for assemblies. The corridors that link the classrooms run around the edge of this courtyard, but they widen and narrow as they do so, creating a changing, irregular space, more amenable for stopping and gathering than the straighter, narrower standard version. The irregular plates that make up the mountain-landscape roof continue to jut out over the playground at odd angles, creating a variety of different spaces around the edge. The richness of this interior space is what helps to break down this institution into something where individuals might more easily find their own place.

Giancarlo Mazzanti's Pies Descalzos School in Cartagena, Colombia is bigger again – it's a major building for its neighborhood. It's a pre-school, middle and high school all in one, along with the ancillary library, halls and sports facilities that such a large school needs. Each of the three individual schools is organized around its own hexagonal courtyard, a shaded space with planting to encourage a cooler micro-climate. The size of the institution and the repetition of forms within it might have made this quite an impersonal building and yet it has a real playfulness, such as the ramps that snake up around the courtyard, ideal for games and running around. The science labs cantilevered out from the main building, with their large windows taking in the view, and connecting back to the local area and wider landscape.

农场幼儿园，同奈省，越南
Farming Kindergarten, Dongnai, Vietnam

位于哥伦比亚考卡省比亚里卡镇、由Daniel Joseph Feldman Mowerman和Iván Dario Quiñones Sanchez设计的El Guadual儿童早期发展中心，是一个集教育和医疗于一体的设施。它将建筑和布局的很多方面联系起来，以研究如何建造友好的（而非给人疏远感的）公共机构。作为学校，它大约可以为三百名幼儿服务，同时，又可以为孕妇和新生儿提供医疗护理。因其能够提供多种服务，所以对比亚里卡镇的人们来说它又是一个重要的社区中心。

像其他学校一样，El Guadual中心也围绕着一个中心庭院而建，庭院为孩子们提供了安全的玩耍场所，同时也为访客提供了一定程度的隐私空间。建筑内部包含十间教室、一个餐厅、厨房、菜园、艺术空间、一座公共剧院以及市民广场等，它们都集中在这处场地内，为附近街区提供了很多服务，这一系列服务给社区带来的生机与活力，是一座单一功能型建筑（即便建筑内全是活泼的孩子们）所无法拥有的。另外，这一设计是通过与本地社区的人们进行一系列的专家研讨会后研究出来的，这也是至关重要的一点。这意味着这一中心的设计和功能不是一个自上而下的实施过程（20世纪大型公共设施的一个重要特点），而是一个容许自下而上发展的过程。最后值得一提的是，该中心所体现的哲学理念可以清晰地从其建筑风格中看出来。该中心支持"雷焦·艾米利亚"教学系统的教学理念，这一理念认为孩子们必须在其本身学习中扮演领导者的角色。而这其中的一个重要因素就是处理好与他人以及当地环境之间的关系。建筑师们满腔热情地接受了这个挑战，在不同的空间之间建立多条通道来解决这一问题。这些通道不仅有我们可能期望的径直的道路，同时也有建造的山岭、桥梁、楼梯以及滑梯等。这样每个学生都可以选择他们自己喜欢的方式在建筑内穿梭。这种赋予个人选择权利的方式也是一种重要方法，以削弱医疗和教育建筑的公共机构特征。如果El Guadual中心的大胆路线是一次合适的娱乐性尝试，那么它也可能是一个极端的事情，但是毫无疑问，这种自下而上的设计过程帮助了一个社区接受这种类型的建筑。建筑在其中扮演着重要角色，为个人用户提供可以自由支配的隐私和空间，并且氛围明亮、有趣。最后一项要点是，削弱那些长长的走廊带来的影响，因为这些走廊是将组成建筑物所必需的众多房间联系在一起的元素。

The El Guadual Children Center by Daniel Joseph Feldman Mowerman and Iván Dario Quiñones Sanchez in Villarrica in the Cauca department of Colombia is a combined education and health facility and it brings together many of the lessons, both architectural and organizational, about how to create an institution that welcomes rather than alienates. It functions as a school for around 300 young children, but also offers medical care to pregnant women and newborn babies. The mix of services on offer means that it is an important community center for the people of Villarica.

Like the other schools, El Guadual is arranged around a central courtyard, where children can play safely and which offers a degree of privacy to visitors. There are ten classrooms, a dining hall, kitchen, vegetable garden, arts spaces, a public theater and civic plaza all crammed onto the site. It offers a lot to the nearby neighborhood and this range of activity brings a liveliness that a more mono-functional building, even one full of children, might lack. Crucially, the design was also developed in close collaboration with the local community through a series of charettes. That meant that the design and program of the center weren't a top down imposition – also a key feature of large 20th century institutions – but instead were allowed to develop from the bottom up. The final thing to mention is the center's philosophy, something that can clearly be seen in the architecture. The education model that the center espouses is the "Reggio Emilia" pedagogic system, which holds that children must take a leading role in directing their own learning. A key element of this is their relationships with others and with the local environment. The architects took up this challenge enthusiastically, offering multiple routes between different spaces, not only via the straightforward paths we might expect, but also through creating mountains, bridges, stairs and slides. Individual child can determine his own route through the building.

It is this empowering of individuals that seems to be key to soften the more institutional aspects of health and educational buildings. The adventurous routing of El Guadual may be an extreme, if appropriately playful, example, but the bottom up design process also no doubt helps in getting a community to fully take on a building of this type. Architecture has a key role to play in this, creating privacy and spaces that can be taken over fully by their individual users, adding a playful lightness and last but not least, mitigating the impact of all those long corridors that connect the many rooms that necessarily make these buildings up. Alison Killing

Råå日托中心

Dorte Mandrup Arkitekter

　　Råå Forskola幼儿园坐落在风景优美的海滩上,位于老Råå学校和大海/厄勒海峡之间。项目依托于周围的景观而建,景观还包括平坦的、带有缓缓坡度的沙丘以及小型渔民住宅所形成的不同地形。

　　该机构可以容纳四组儿童,每组儿童都有自己的房间。这四间小组房间的立面设计与渔民住宅的山墙一致。公共空间则位于山墙之间。立面上的大型窗户和屋顶不但与大海和周围景观建立了密切的联系,而且全年都提供了良好的日照条件。

　　在视觉方面,小组房间被胶合木制成的"书架墙"所围合。它们连同绿色的地面一起形成了流动的空间体验,且整体一致性和透明性贯穿整个机构。该机构与原有的老学校相连接,连接处设有一处洞穴式空间,用作衣柜。操场面向老学校,且受到了大海/厄勒海峡的庇护。面向大海/厄勒海峡的海滩上的自然物种重新进行了栽植。该项目已经运作很长时间了,从2008年开始,便与未来的用户进行密切的对话。

Råå Day Care Center

Råå Forskola, a kindergarten, is situated on the scenic beach between the old Råå School and the Sea/Øresund. The building is based on the surrounding landscape, with its flat slightly sloping dunes and the distinctive typology of the small fishermen houses. The institution comprises four groups of children, each with its own group room. The four group rooms identify themselves to

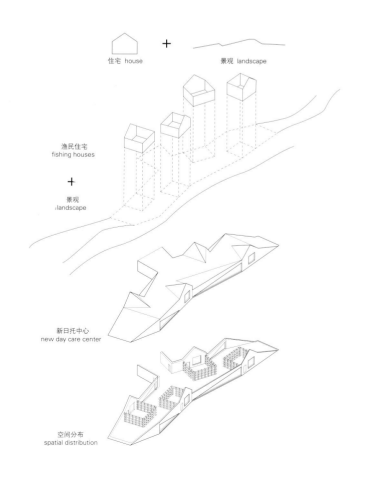

the facade as the gable of a fisherman house. Located between the gables are the common spaces. Large windows in the facade and roof create a close contact with the sea and the surrounding landscape, and provide ideal daylight conditions all year round. The group rooms are visually enclosed by "bookcase-walls" made of plywood. Together with the green floor they create a fluent spatial experience with consistency and transparency through the entire institution. The institution is linked to the existing school and in the connection a cave-like space for wardrobes is created. Sheltered from the Sea/Øresund toward the old school, a playground is situated. Facing the Sea/Øresund the natural flora of the beach has been replanted. The project has been a long time coming, beginning in 2008, in close dialogue with the future users.

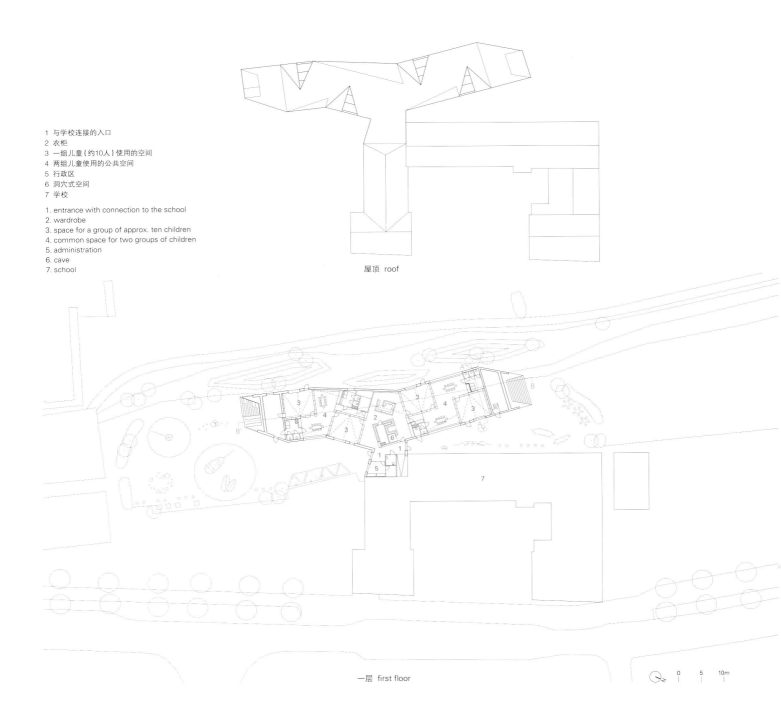

1 与学校连接的入口
2 衣柜
3 一组儿童（约10人）使用的空间
4 两组儿童使用的公共空间
5 行政区
6 洞穴式空间
7 学校

1. entrance with connection to the school
2. wardrobe
3. space for a group of approx. ten children
4. common space for two groups of children
5. administration
6. cave
7. school

屋顶 roof

一层 first floor

1 一组儿童（约10人）使用的空间 2 浴室
1. space for a group of approx. ten children 2. bathroom
A-A' 剖面图 section A-A'

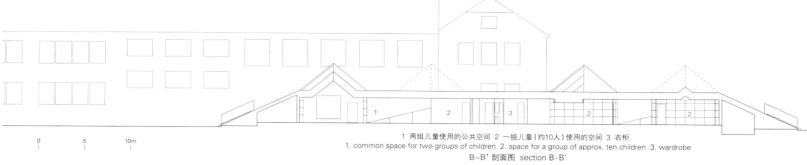

1 两组儿童使用的公共空间 2 一组儿童（约10人）使用的空间 3 衣柜
1. common space for two groups of children 2. space for a group of approx. ten children 3. wardrobe
B-B'剖面图 section B-B'

项目名称：Råå Day Care Center
地点：Kustgatan 1, 252 70 Råå, Sweden
建筑师：Dorte Mandrup Arkitekter
施工工程师：Tyréns AB
能源工程师：Ramböll Sverige AB
景观建筑师：Marklaget AB
甲方：City of Helsingborg
用地面积：525m²
设计时间：2008
竣工时间：2013.8
摄影师：©Adam Mørk (courtesy of the architect)

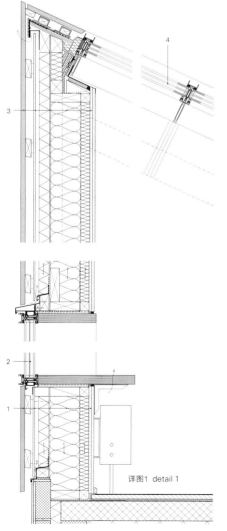

详图1 detail 1

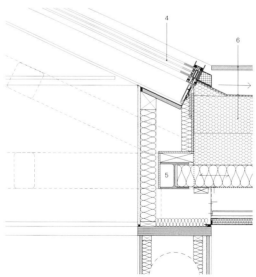

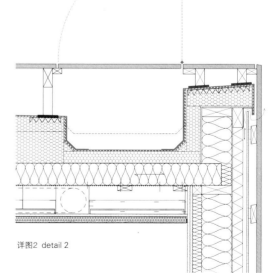

详图2 detail 2

1. 22x40mm, pre-weathered robinia moldings, distance: 8mm
   "regular" 8mm, cement fibreboard, black
   50 mm, hard insulation (for the climate)
   190mm, insulation
   2x13mm, gypsum fibreboard
2. aluminum frame with triple glazing, two are safety glass
   the indoor frame is 45mm kerto
3. 22x40mm, pre-weathered robinia moldings, distance: 8mm, metal studs
   8mm, cement fibreboard, black
   50mm, hard insulation (for the climate)
   190mm, insulation
   2x13mm, gypsum fibreboard
4. self-supporting aluminium frame
   roof light with triple glazing, two are safety glass
5. kerto space-dividing racking systems and steel beams
6. 22x40mm. pre-weathered robinia moldings, distance: 8mm,
   lath build-up
   black roofing cardboard
   foam glass
   wood-wool suspended acoustic soffit

社区建筑——个性化建筑 In the Community—The Individuality and the Institution

农场幼儿园
Vo Trong Nghia Architects

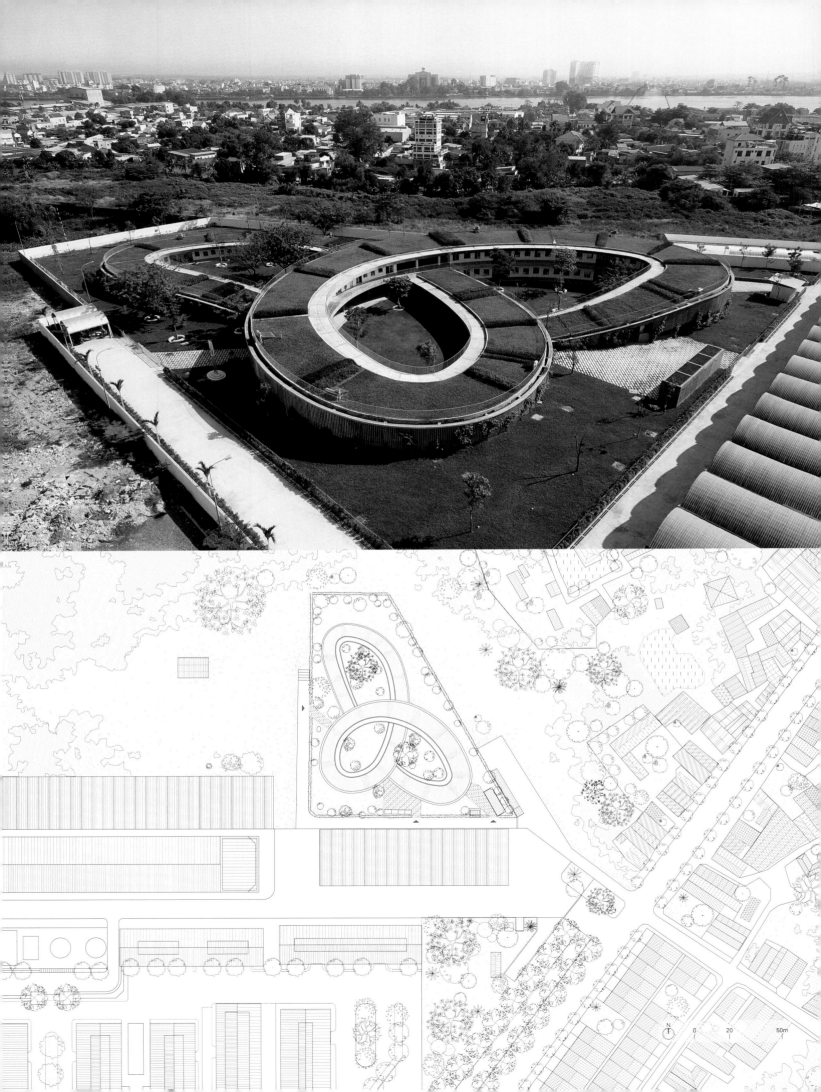

项目名称：Farming Kindergarten / 地点：Dongnai, Vietnam
主要建筑师：Vo Trong Nghia, Takashi Niwa, Masaaki Iwamoto
建筑师：Tran Thi Hang, Kuniko Onishi
承包商：Wind and Water House JSC
绿色建筑顾问：Melissa Merryweather
CFD分析：Environment Simulation Inc.
甲方：Pou Chen Vietnam
用地面积：10,650m² / 有效楼层面积：3,800m² / 竣工时间：2013.10
摄影师：
©Hiroyuki Oki (courtesy of the architect) - p.118~119,
p.120, p.121, p.122, p.123
©Gremsy (courtesy of the architect) - p.114~115, p.116

这座幼儿园有500个学前儿童，位于一个大型鞋厂附近，是在热带气候地带建造可持续教育空间的原型。这座建筑的预算较低，是为工厂工人的孩子设计的。

越南有许多用于农业生产的土地，例如湄公河三角洲。但是最近，这些土地出现了许多问题，它们遭受了自然和农业的侵蚀，如洪水、盐碱地和干旱等。

同时，越南也有很严重的城市问题。大量的摩托车在街道穿行，导致空气污染。因此，越南的城市失去了绿地。越南没有给孩子提供玩耍的安全地带，孩子们变得很不活跃。针对这些问题，建筑师的任务是为越南的孩子建造一座绿色的幼儿园。

该建筑的设计概念是农场幼儿园，通过建立一个连续的绿色屋顶，为越南的孩子们提供粮食和农耕体验，同时也提供一处安全的户外活动场所。三个环形的绿色屋顶，像画笔一挥，营造出三个内部庭院，这些内部庭院为孩子们提供了安全且舒适的游乐场地，使孩子们在两侧下到庭院，当孩子们在其中上下穿梭游玩时，能拥有和自然生态和谐相处的特殊经历。绿色的屋顶成为蔬菜园，教会孩子们农业的重要性以及与自然之间的亲密关系。

综合运用建筑和机械的节能方法包括如下方式（但不仅限于此）：绿化屋顶、预应力混凝土遮阳百叶、回收的材料、循环用水系统、太阳能供暖设备等。这些直观的设计是为了让孩子们意识到可持续发展教育的重要性。

Farming Kindergarten

The Kindergarten for 500 preschool children, situated next to a big shoe-factory, is a prototype of the sustainable education space in tropical climate. The building is designed for the children of factory workers within low-budget.

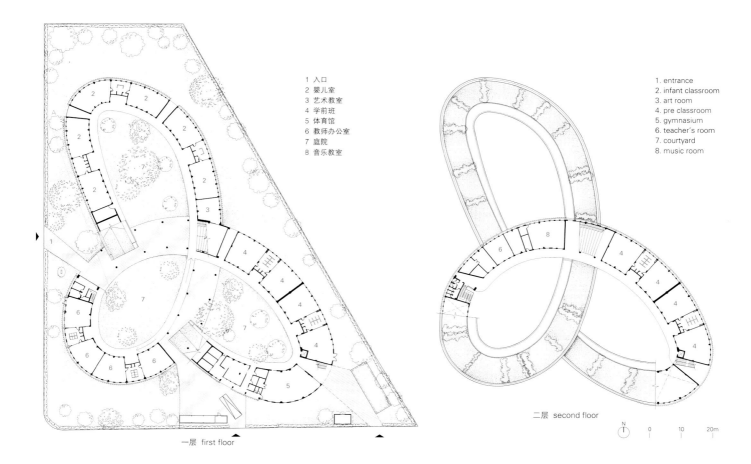

1. 入口
2. 婴儿室
3. 艺术教室
4. 学前班
5. 体育馆
6. 教师办公室
7. 庭院
8. 音乐教室

一层 first floor

1. entrance
2. infant classroom
3. art room
4. pre classroom
5. gymnasium
6. teacher's room
7. courtyard
8. music room

二层 second floor

Vietnam has many productive lands for agriculture, for example Mekong Delta. However, recently, there are many problems in this area, which suffers nature and agriculture, such as flood, salt damage and drought.

There are serious urban issues, too. Numerous motorbikes are running in streets, causing air pollution. And cities in Vietnam lost its greenery. Consequently, there is no safe playground for children, and children is getting inactive. The architects' mission is to creat a green kindergarten for Vietnamese children against these problems.

The concept of building is "Farming Kindergarten" with continuous green roof, providing food and agriculture experience to Vietnamese children, as well as safe outdoor playground. The green roof is triple-ring-shape drawn with a single stroke, creating three

连续的剖面 continuous section

courtyards inside. While these internal courtyards provide safety and comfortable playgrounds for children, the roof makes landing to the courtyards at both sides, allowing children to enter a very special eco-friendly experience when they walk up and go through it. This green roof is designed as a vegetable garden, a place to teach children the importance of agriculture and relationship with nature.

Architectural and mechanical energy-saving methods are comprehensively applied including but not limited to: green roof, PC-concrete louver for shading, using recycle materials, water recycling, solar water heating and so on. These devices are designed visibly in order for children to realize the important role of sustainable education. Vo Trong Nghia Architects

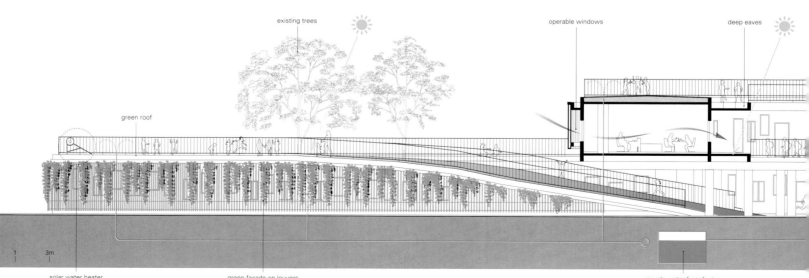

A-A' 剖面图 section A-A'

Pies Descalzos学校

Mazzanti Arquitectos

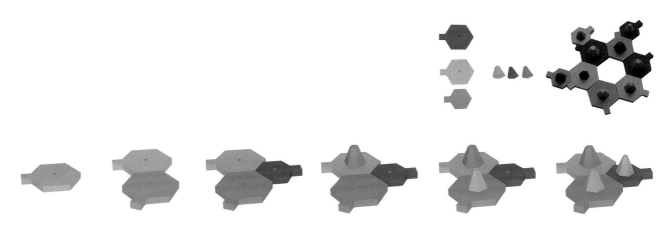

这座学校位于哥伦比亚卡塔赫纳德印第亚斯内，Loma del Peye山的顶端，是为Pies Descalzos基金会而设计的。这个项目不仅是一座教育性建筑，还试图通过给个人和社区发展提供多种选择，改变这个地区的环境来唤起居民的团结，提高生活质量。这个项目已经成为城市和农业的地标以及整个社区都感到自豪的标志。

这座建筑建在一系列三个横切的六边形结构上。每个六边形结构界定了两层结构的边界，中间设有庭院。六边形的外部轮廓还定义了边界的流线，且容纳了学校的功能项目。另一方面，覆顶的庭院种植了各种树木和其他植物，形成微气候环境，为不同性质的活动提供了独有的特色和布局。庭院里选择栽种的植物有助于吸引当地野生动物，鼓励生态教育的发展。这座建筑是一处有归属感、明亮且安宁的地方。

这座建筑通过创造一个能够全面利用场地的水平序列，来适应当地的地形。学前区独立于建筑的其他部分，成为一个独立的楼层。图书馆位于一层，因为它对开放性有特殊需求，且要满足公众从街上直接进入的需要，具有开放性。中学和高中区位于两个大六边形结构内。楼层通过中央的斜坡和楼梯来进行垂直连接。建筑突出来的结构是科学实验室，设有大型窗户，可以从视觉方面与城市相连接。另外，学校有多功能厅，带有一个体育设施，面向一个入口广场开放。

Pies Descalzos School

This school, designed for Pies Descalzos Foundation is located on the top of Loma del Peye Mountain within the city of Cartagena de Indias, Colombia. More than an educational building, the project seeks to provoke the consolidation of the neighborhood and improve the life of residents by generating alternatives for personal and community development as well as an environmental transformation of the area. The project has already become an architectural and urban landmark and an object of pride for the community.

The design of the building is based on a sequence of three intersected hexagons. Each hexagon defines a two level perimeter with a central patio. The hexagon contours define a perimeter cir-

1 多功能室	16 学校商店
2 体育场更衣室	17 广播室
3 存储室	18 儿童教室
4 化妆室	19 幼儿园
5 卫生间	20 心理咨询室
6 废物回收和垃圾中心	21 小组工作室
7 清洁室	22 艺术教室
8 应急备用设备	23 家长室
9 配电站	24 校长办公室
10 洗衣房	25 图书馆
11 存储室和工作间	26 机房
12 教室	27 礼堂
13 厨房	28 教师咖啡室
14 急诊室	29 教师办公室
15 音乐教室	

1. multiple use room	16. school store
2. sports changing room	17. radio
3. storage	18. children's room
4. fitting room	19. kindergarten
5. w.c.	20. psychology
6. recycling and trash center	21. group work room
7. cleaning	22. art room
8. emergency plant	23. parents room
9. electrical substation	24. principal's office
10. laundry	25. library
11. storage and workshop	26. computer room
12. classroom	27. auditorium
13. kitchen	28. teachers coffee room
14. infirmary	29. teachers room
15. music room	

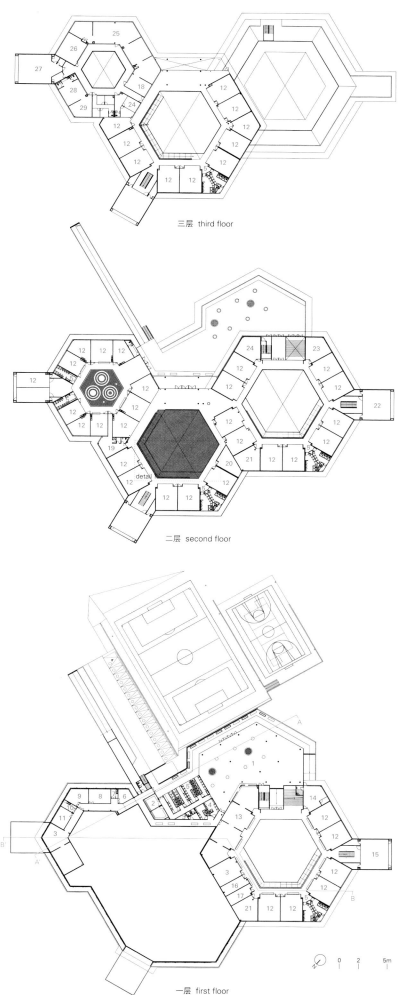

三层 third floor

二层 second floor

一层 first floor

项目名称：Pies Descalzos School
地点：Loma del Peye, DTC Cartagena de Indias, Colombia
建筑师：Giancarlo Mazzanti
项目经理：Juan Andres Lemus / 副经理：Juan Manuel Gil
草图绘制：Nestor Gualteros, Rocio Lamprey
项目开发：Fredy Fortich, Rocio Lamprey, Liv Johana Zea,
Diego Castro, Maria Sol Echeverri
实习生：Stephany Zapata, Juliana Alzate, Andres Correa,
Jessica Jaramillo, Samir Abu-Shibab, Santiago Rincon, Fabvio Caicedo,
Alejandra Loreto, Christelle Tsikras, Manuel Gutierrez
结构工程师：Nicolás Parra
甲方：Barefoot Foundation
总建筑面积：11,200m² / 有效楼层面积：14,000m²
设计时间：2011 / 施工时间：2013 / 竣工时间：2014.2
摄影师：©Sergio Gomez (courtesy of the architect) - p.128~129,
p.130^{bottom}, p.132~133, p.135
©Pies Descalzos Foundation (courtesy of the architect) - p.126~127, p.130^{top}

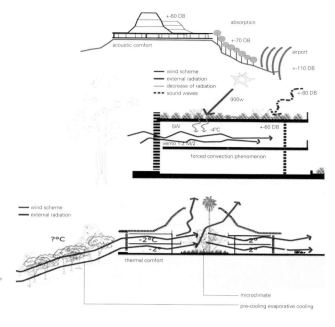

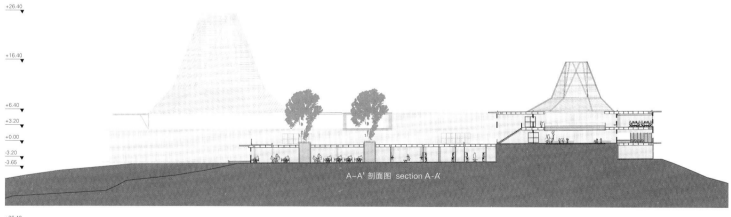

A-A' 剖面图 section A-A'

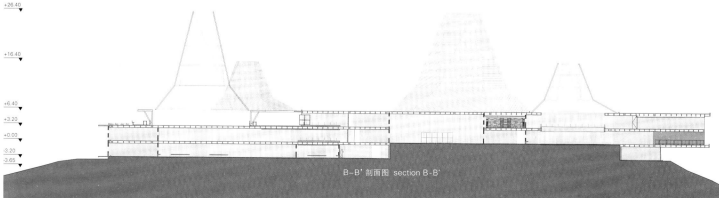

B-B' 剖面图 section B-B'

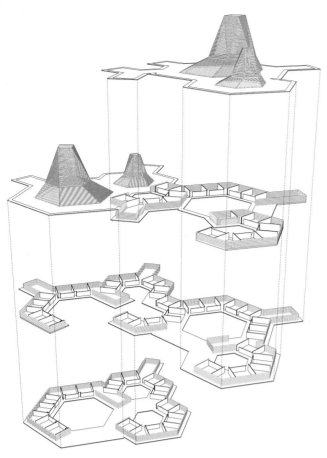

culation and contain the school's program. On the other hand, the covered patios have a variety of trees and plant species that create a micro-weather, giving a specific character and disposition for different activities to take place. The vegetation selected will also attract the presence of native animals, encouraging the development of an ecological education. The architecture is a space of belonging, light and calm.

The building adapts to the geography of the place by creating a level sequence that permits the full use of the area. The preschool area is independent from the rest of the building and functions in one level. The library also functions on the first floor as it needs special conditions of access as well as direct and public access from the street. The middle and high school areas are located in two bigger hexagons. Levels are connected vertically by a central ramp and stairs. The elements that come out from the building are science labs that have big windows to provide a visual connection to the city. Additionally, the school has a multiple use hall, with sport services opened to an access plaza.

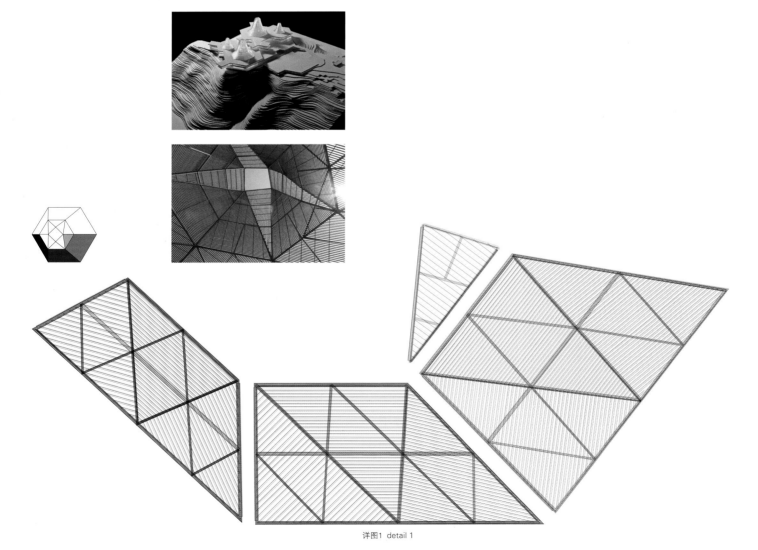

详图1 detail 1

Zugliano的某所学校

5+1AA Alfonso Femia Gianluca Peluffo

该建筑与外界明显隔离开来。其"壳状"覆层代表农业建筑、市政公共市场和18世纪启蒙运动时期建筑传统中公共建筑具有代表性和纪念性的理念之间的一种中间产物。这种几乎由传统的陶瓷涂层覆盖的、由半遮盖式地基组成的半角锥和半圆锥体结构将成为Zugliano社区一个易于识别的独特元素。

建筑师可以想象一栋既具代表性又具备功能性的建筑,它简单且易于识别,能够与所在领域和景观进行对话,既受到保护又具备防护性。该项目基于两种地域性元素:地块的水平布局、其清晰度和"简单性";山脉和丘陵的轮廓。项目的主题是利用这些条件,来建造一栋既具有代表性同时又具备功能性的建筑。

对当地传统和对一栋独立建筑主题的解读引导建筑追求简单的几何/结构设计,使之能够成为"场地内的建筑,同时塑造场地",也就是说,形成与环境相得益彰,同时又强调其自身特征的系统。

因此,建筑师们在楼层平面设计阶段就决定将简单性、学校设计主题的功能和社会需求(庭院、受保护的开放空间)和历史传统(庭院、农家庭院)结合在一起:形成一栋具有象征性和表现力的大型室内空间的正方形建筑。

山丘和阿尔卑斯山脉的轮廓间进行对话的设计愿景赋予了该建筑帐篷状的外观设计理念,形成一个开放的圆锥体。因此,建筑的屋顶拔地而起,其外观成为这所学校具有特色并且易于识别的元素,并与庭院的开口保持一致,使该结构可见并形成建筑的一颗愉快、多彩、欢快的"跳动心脏",同时"让阳光照进庭院",在建筑内部绘制自身的形态。

在领域规模方面,该建筑具有特色的建筑构件是覆层,带有不同的坡度,这些坡度的角度略有不同。整栋建筑的特征在于一条受保护的小径,即内部庭院周围的一个特有的门廊。该门廊连接整个学校系统,形成自由空间,并且具有学校独有的分段时间的特征。

教室的面积几乎都为45m²,根据规定,每间教室将容纳25名学生。这些教室的特点在于带有连续的洞口(始于60cm的内部高度)设计,同样,学生在坐位就可以欣赏到外部的景色;覆层面向内部庭院倾斜,以表现方向的重要性,以及公共领域和建筑作为一个整体的社区想要表达的理念:在建筑传统中,门廊-庭院系统是一个连接会客和沉思主题的公共社区系统。

School Complex in Zugliano

Apparently closed to the outside, this "shell"-covering represents a kind of intermediate element between the farm buildings, the municipal public markets and the iconic and monumental idea of public building of the eighteenth century, belonging to the tradition of Enlightenment architecture. This semi-pyramid and semi-cone, consisting of an almost traditional ceramic-coated, semi-glossed base, will become a recognisable and distinctive element of the community of Zugliano.

The architects imagined a building that was both representative and functional, recognisable and simple, with the ability to communicate with the territory and the landscape, but protected and protective. The project is based on two territorial elements: the horizontal layout of the lot, its cleanliness and "simplicity"; the profile of the mountains and hills. The theme of the project is to exploit these conditions in order to construct a building that is representative and functional.

The reading of the local traditions and the theme of an isolated building led the architects to pursue the path of geometric/compositional simplicity, capable of "building in the place and build-

ing the place", that is, a system that declares its belonging to the context, strengthening its own characteristics.

Thus, at the level of floor plan, it was decided to combine this push toward simplicity with the functional and social needs of the school theme (the courtyard, protected open space) and a historical tradition (the courtyard, the farmyard): a square building with an interior space, large, symbolic and expressive.

The desire to imagine a dialogue with the outline of the hills and the Alps gave life to an idea of the profile of the building almost as a tent shape, an open pyramid. Thus, the roof of the building, of which outside is the characterising and recognisable element of the school, rises up, and opens in correspondence to the courtyard, in order to make "visible" its structure and declare a joyful, colourful, cheerful "beating heart" of the building, and "to let the heavens in", drawing their form inside the building itself.

The architectonic element that characterises the building to the scale of the territory is the covering, which is made up of different pitches, with slightly variable tilts. The whole building is characterised by a protected path, a proper portico, around the internal courtyard. This portico connects the entire school system, allowing freedom of space and split times that are typical of school.

The classrooms are almost 45 square metres and each, in line with the regulations, is for 25 pupils. They are characterised by a continuous opening, starting from an internal height of 60 cm, which allows a view to the outside also from a sitting position; the covering is tilted up towards the inner court, to mark the importance of that direction, and the sense of "community" that common area and the building as a whole want to express: the portico-courtyard system is, in the tradition of building, a public-community system, linking to the themes of meeting and reflection.

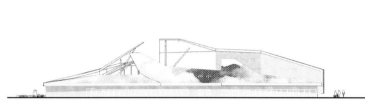
西立面 west elevation

东立面 east elevation

南立面 south elevation

项目名称：School Compex in Zugliano
地点：Zugliano, Vicenza, Italy
建筑师：Alfonso Femia, Gianluca Peluffo
总建筑师：Alfonso Femia, Gianluca Peluffo, Diego Peruzzo, Alessandro Cavaleri
项目经理：Stefania Bracco
合作者：Emanuela Bartolini, Francis Bust, Gianmatteo Ferlin, Helen Graziano, Valentina Grimaldi Fabio Marchiori, Carola Picasso
结构工程师：Iquadro Engineering Ltd.,
协调工程师：Roberto Mancini
服务和环境工程师：Stefano Cremo
总协调员：Simonetta Cenci
甲方：City of Zugliano
用地面积：3,500m² (courtyard 1,000m²)
造价：EUR 4.2m
设计时间：2010
竣工时间：2013
摄影师：©Ernesta Caviolat (courtesy of the architect)

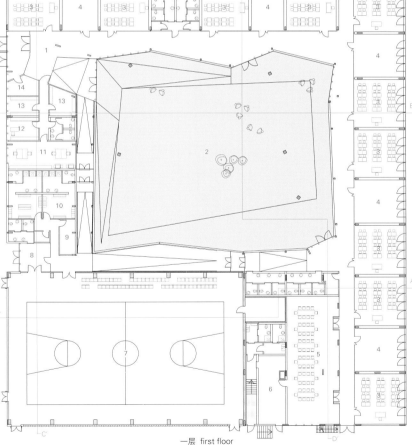
一层 first floor

1 主入口	9 医务室	1. main entrance	9. infirmary
2 庭院	10 更衣室	2. courtyard	10. changing room
3 教室	11 教师公共休息室	3. classroom	11. teachers' common room
4 多功能区域	12 办公室	4. multipurpose area	12. office
5 食堂	13 档案室	5. canteen	13. archive
6 杂物房	14 门房	6. utility room	14. porter's lodge
7 体育馆	15 机房	7. gym	15. plant room
8 次入口		8. secondary entrance	

地下一层 first floor below ground

1. roof consisting of prefa "falzonal" seamed aluminium in different colours, polyethylene sheath, double waterproofing membrane, 5/8" (15mm) osb panel, double frame of 4 3/4 x 3 1/8" (120 x 80mm) wood battens sandwiching rockwool insulation, vapour barrier, 2 3/8" (60mm) timber panel, structural frame of 21 5/8 x 6 1/4" (550 x 160mm) fir glulam beams
2. metal flashing
3. rainwater guttering and drainage system
4. secondary frame of 16 1/2 x 6 1/4" (420 x 160mm) glulam beams
5. floor-to-ceiling continuous glazed facade with 5/16-5/8-1/4-5/8-1/8+1/8" (8/15/6/15/4+4mm) aluminium double glazing units in aluminium frame
6. interior wall consisting of double 1" (25mm) gypsum board, double supporting frame of 2 3/8" x 1 1/4" (60 x 30mm) aluminium c-profiles, 2 3/8" (60mm) rigid insulation, double 1" (25mm) gypsum board
7. false ceiling consisting of 23 5/8 x 23 5/8" (600 x 600mm) natural fibre panels on frame of 2 3/8 x 1 1/4" (60 x 30mm) aluminium c-profiles suspended from slab by tie rods
8. partition wall with aluminium frame and 1/4 + 1/4" (6+6mm) safety glass
9. floor comprising dry-pressed ceramic tiles, 2 7/8" (75mm) screed over radiant heating, 1 1/4" (30mm) rigid insulation, 3 1/8" (80mm) reinforced concrete slab, vapour barrier, 4" (100mm) rigid insulation, reinforced concrete slab with plastic void formers
10. paving comprising 1 1/4" (30mm) polished concrete, 3 3/4" (95mm) max h screed forming slope, 4" (100mm) reinforced concrete slab, gravel
11. branched column formed by 15 3/4 x 15 3/4" (400 x 400mm) glulam column (parallel to plane of section)
12. rope finish on base of column

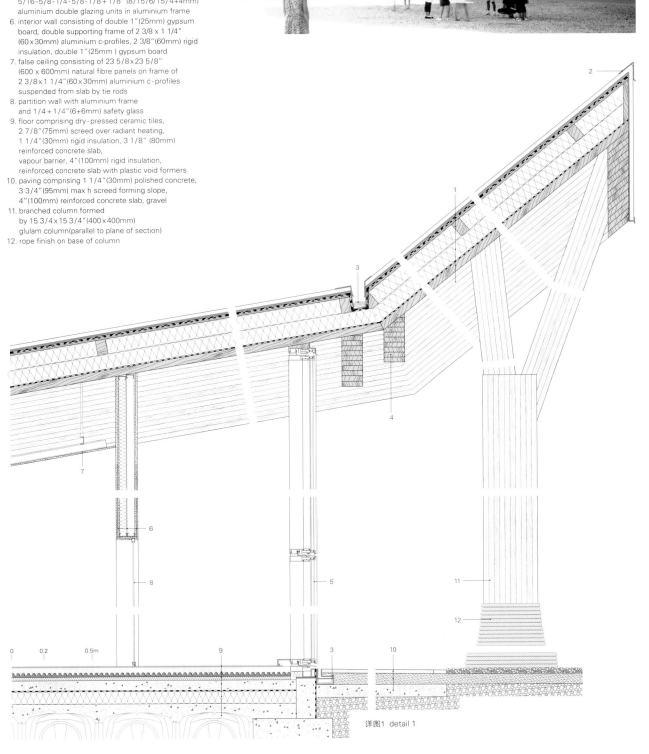

详图1 detail 1

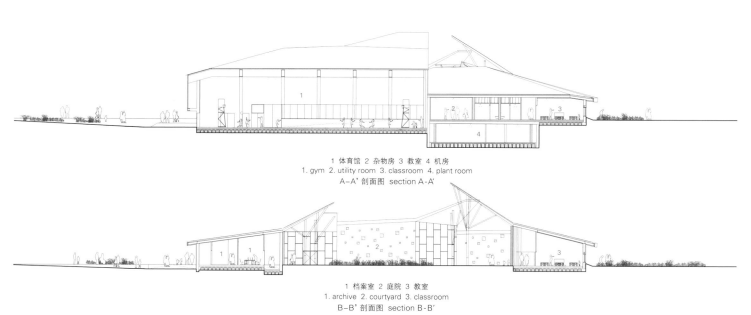

1 体育馆 2 杂物房 3 教室 4 机房
1. gym 2. utility room 3. classroom 4. plant room
A-A' 剖面图 section A-A'

1 档案室 2 庭院 3 教室
1. archive 2. courtyard 3. classroom
B-B' 剖面图 section B-B'

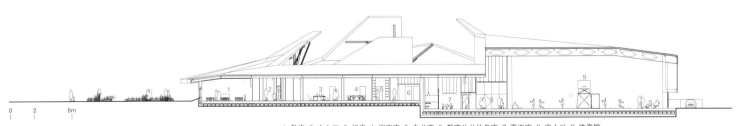

1 教室 2 主入口 3 门房 4 档案室 5 办公室 6 教室公共休息室 7 更衣室 8 次入口 9 体育馆
1. classroom 2. main entrance 3. porters' lodge 4. archive 5. office 6. teachers' common room 7. changing room 8. secondary entrance 9. gym
C-C' 剖面图 section C-C'

1 教室 2 庭院 3 食堂 4 机房
1. classroom 2. courtyard 3. canteen 4. plant room
D-D' 剖面图 section D-D'

El Guadual儿童早期发展中心位于考卡的Villa Rica，包含十间教室、餐厅、室内和户外休闲区、半私人的艺术空间、急救室、行政办公室、菜园、水景、室外公共剧院以及一个市民广场，且作为国家整体青年战略"de Cero a Siempre"的一部分，能够为300名0到5岁的儿童、100名孕妇以及200名新生儿提供食物、教育和休闲服务。

该中心于2013年十月举办竣工典礼，标志着为期三年的参与式设计和施工工作结束。其间人们始终坚持不懈地为他们的自豪感与归属感而不断努力。与当地孩子、青少年、雇员和社区领导人一起玩设计猜字游戏是整个项目在空间、材料、尺寸、与城市关系方面的出发点。整个施工过程一共持续了九个月，总造价为160万美元。建造这所学校的资金来自国际合作、私人捐赠、公共集资以及以货代款的捐赠。整个项目一共聘用了60多名合格的当地施工人员以及30名早期在青年教育中接受过培训的合格女工，他们是该中心的劳动力。

El Guadual儿童早期发展中心因其为公众提供了大量的人行道和景观，并设有一座开放的室外公共影院、一间半私人的艺术和表演室（在晚上和周末向社区开放）以及一个市民广场，而对城市产生了显著的影响。一排的公共设施使该中心成为Villa Rica内一处全新的活动区。

采用雷焦·艾米利亚教学系统的十间教室提供了开放的空间、障碍物以及多变的因素，中心使这个发现其本身的过程变得充满挑战与乐趣，教学成为休闲式的体验。通过山丘、桥梁、楼梯以及滑动门窗，若干个相连接的入口和出口将教室进行了配对，从而通过建筑物营造一种促进决断思维以及个人发展的氛围。每间教室都配备了一间浴室，使孩子们在任何时候，而不仅仅是在教师专注于教学活动时，都可以使用。

整个项目是一个典型的低技术环保型建筑，这与使用的材料及其耐久性、水源和消耗的能源相关。所有的空间都能全天接收到光照，且通风良好，使该中心无需能源也可正常运行。带有纹理的混凝土墙体能够吸收热量，从而保持室内空间的凉爽。而多层的屋顶能够控制阳光对室内空间造成的影响。建筑师们采用当代的方法，使竹子的使用成为重新审视当地传统的方式，不但充分利用了当地元素，而且还对河床起到了保护作用。每间教室都能收集雨水，雨水可以用于园艺和日常维护，而且孩子和游客也能有机会看见雨水收集和利用的过程。中央的水景不断地再循环，孩子们能够与其进行互动，使水又变成了一个娱乐性的元素。带有纹理的墙体采用当地的技术，由劈开的竹子模板制成。在该中心照顾孩子们的老师又将他们收集的废旧瓶子扣在栅栏的顶端。

El Guadual儿童早期发展中心

Daniel Feldman & Iván Dario Quiñones

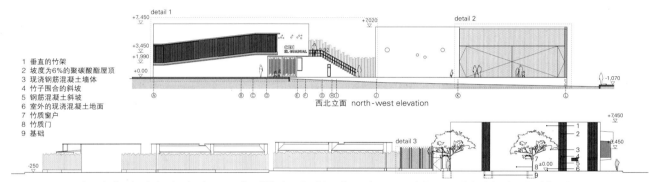

1 垂直的竹架
2 坡度为6%的聚碳酸酯屋顶
3 现浇钢筋混凝土墙体
4 竹子围合的斜坡
5 钢筋混凝土斜坡
6 室外的现浇混凝土地面
7 竹质窗户
8 竹质门
9 基础

西北立面 north-west elevation

1. vertical bamboo lattice 2. polycarbonate roofing of 6% slope 3. reinforced poured concrete wall 4. ramp enclosing in bamboo 5. reinforced concrete ramp 6. external floors in poured concrete 7. bamboo window 8. bamboo door 9. root container

东北立面 north-east elevation

西南立面 south-west elevation

1 现浇钢筋混凝土墙体
2 坡度为6%的聚碳酸酯屋顶
3 垂直的竹架（直径为5cm）
4 钢筋混凝土墙体上的钢筋混凝土斜形坡面
5 带有纹理的混凝土地面
6 面板之间直径为7cm的空隙

1. reinforced poured concrete wall
2. polycarbonate roofing of 6% slope
3. vertical bamboo lattice ø5cm
4. reinforced concrete ramp anchored to concrete wall
5. textured concrete floors
6. 7cm spacing between slabs

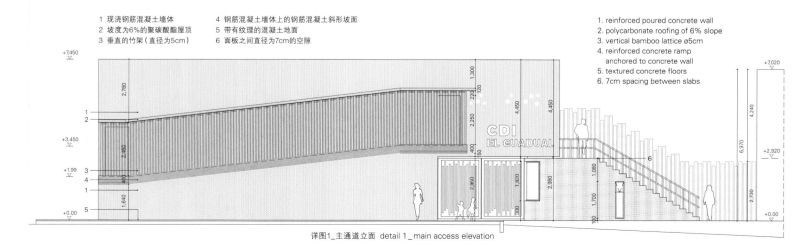

详图1_主通道立面 detail 1_main access elevation

最后，El Guadual儿童早期发展中心正在逐渐地转变成一个城市中心，在这里，教育活动和艺术活动在进行，且几代人聚集在一起，共同使青少年早些关注社会责任。

El Guadual Early Childhood Development Center

Composed of 10 classrooms, dining hall, indoor and outdoor recreation, semiprivate arts spaces, first aid room, administration, vegetable garden, water feature, public outdoor theater, and a civic plaza, El Guadual Early Youth Development Center in Villa Rica, Cauca provides food, education, and recreation services to 300 kids 0-5 years old, 100 pregnant mothers, and 200 newborns as part of the National Integral Early Youth Attention Strategy "de Cero a Siempre".

The Center's inauguration in October 2013, marked the end of a three year long participatory design and construction effort that has strived to generate pride and ownership since the beginning of the process. Design charades with local kids, teenagers, early youth workers, and leaders were the starting point of the design in terms of spaces, materials, dimensions, and relations with the city. The construction lasted 9 months and the total cost of the project was USD 1.6 Million. The funds to build the project came from international cooperation, private donations, public resources, and in kind donations. During the construction process more than 60 local builders were employed and certified in construction techniques. Additional to the construction jobs, 30 local women were trained in early youth educator before being certified and hired to become the daily workforce of the center.

El Guadual has generated a notable urban impact as it offers generous sidewalks and landscape to the public, an open public outdoor movie theater, a semiprivate arts and performing room

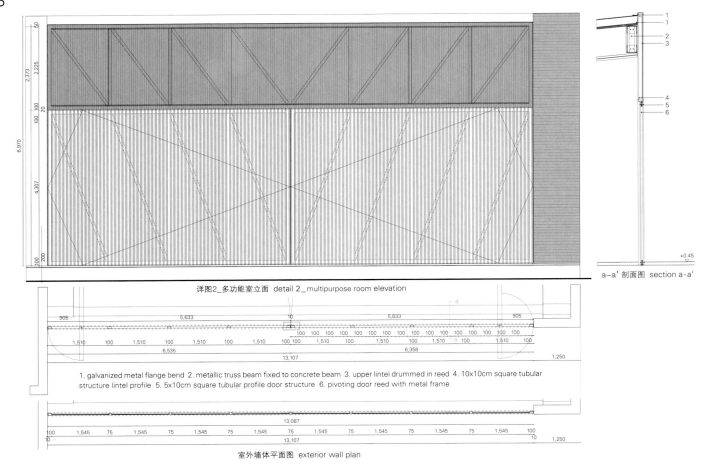

详图2_多功能室立面 detail 2_multipurpose room elevation

1. galvanized metal flange bend 2. metallic truss beam fixed to concrete beam 3. upper lintel drummed in reed 4. 10x10cm square tubular structure lintel profile 5. 5x10cm square tubular profile door structure 6. pivoting door reed with metal frame

室外墙体平面图 exterior wall plan

a-a' 剖面图 section a-a'

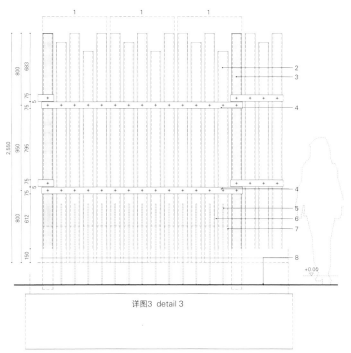

1. typical segment
2. bamboo fence ø11cms
3. columns, round tubular galvanized steel ø11cms, rust proof paint, embedded in concrete pedestal
4. iron mooring 3"x 1/4"
5. internal reinforcement rod anchor
6. reinforced concrete drainage pedestal
7. first cavity of bamboo filled with concrete
8. exterior ground line
9. top to do with the community
10. anchorage stick
11. first cell fill in cement
12. drain 2" in concrete beam c./2.5cm
13. drain 3" embedded in beam
14. reinforced concrete fescue
15. reinforced concrete beam
16. loose gravel drain
17. common drain pipe ø3" or 4"

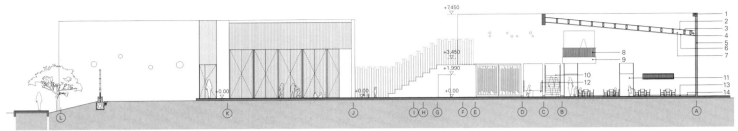

A-A' 剖面图 section A-A'

1. reinforced concrete beam
2. roof 10% slope
3. metal truss structure
4. metal sheet rain water gutter
5. reinforced poured concrete wall
6. bamboo ceiling
7. thermo-acoustic insulation black theater 3"
8. bamboo verticals ø5cm
9. concrete slab
10. reinforced concrete stair
11. hollow wood door
12. reinforced concrete ramp
13. poured concrete floor, polished and sealed
14. curved baseboard in polished and sealed concrete

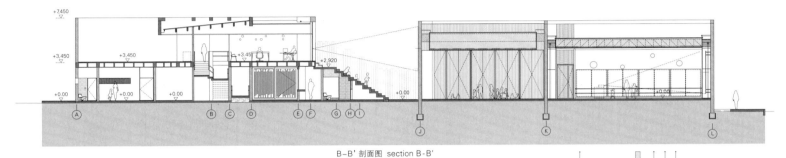

B-B' 剖面图 section B-B'

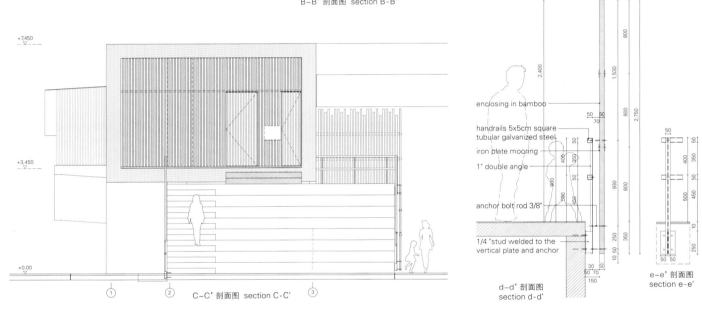

C-C' 剖面图 section C-C'

enclosing in bamboo
handrails 5x5cm square tubular galvanized steel
iron plate mooring
1" double angle
anchor bolt rod 3/8"
1/4" stud welded to the vertical plate and anchor

d-d' 剖面图 section d-d'

e-e' 剖面图 section e-e'

that opens to the community at night and weekends, and a civic square. The array of public amenities has made El Guadual a new pole of activity within Villa Rica.

The 10 classrooms under the Reggio Emilia pedagogic system offer open spaces, obstacles, and multiple variables to navigate the center making the process of discovering the center itself both a challenge and a game making education a recreational experience. Numerous entrances and exits connecting paired up classrooms through mountains, bridges, stairs, and slides foster an environment of decision-taking and individual development through architecture. Each classroom has its own bathroom allowing kids to use it whenever they feel like it, not when the teacher can take themselves to focus on the pedagogical activities.

The project is an example of low tech environmental construction. It is responsible with the environment in terms of the materials it uses, the water and energy it consumes, and the durability of the materials. The spaces all receive natural light throughout the days and are ventilated naturally allowing the center to work without the need of energy. The textured concrete walls absorb heat keeping the spaces cool, and the multi-layered roof controls the impact of the sun on the inside the rooms. The use of bamboo as a way of re-valuing local traditions in a contemporary way speaks of the need to use local materials as well as preserve the riverbeds. Each classroom collects rain water that is used for gardening and maintenance, but makes the process of collection and utilization evident for the kids and visitors. The central water feature recirculates the water it uses and allows kids to interact with water as a recreational element. The textured walls were made using local techniques of split bamboo form work. The fence was capped with recycled bottles collected and installed by the educators who now take care of the kids in the center.

Finally, El Guadual is slowly transforming a new city center where education, arts, and multi-generational gatherings are taking place making the care of the municipalities of early youth a communal responsibility.

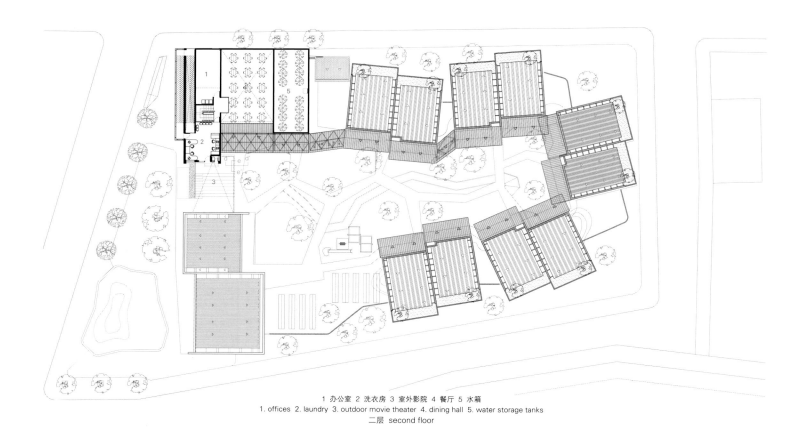

1 办公室 2 洗衣房 3 室外影院 4 餐厅 5 水箱
1. offices 2. laundry 3. outdoor movie theater 4. dining hall 5. water storage tanks
二层 second floor

1 入口 2 餐厅 3 厨房 4 护理区 5 存储室 6 室外影院 7 卫生间 8 室内游乐场 9 多功能室/室内影院 10 舞台 11 日托中心
12 教室（3~23月儿童使用） 13 教室（24~35月儿童使用） 14 教室（36~60月儿童使用） 15 菜园 16 游乐场 17 市民广场
1. entrance 2. dining hall 3. kitchen 4. nursing 5. storage 6. outdoor movie theater 7. w.c. 8. indoor playground 9. multipurpose room / indoor movie theater 10. stage
11. day care center 12. classroom(3 to 23 months) 13. classroom(24 to 35 months) 14. classroom(36 to 60 months) 15. vegetable garden 16. playground 17. civic plaza
一层 first floor

项目名称：Early Childhood Development Center El Guadual
地点：Villa Rica, Cauca
建筑师：Daniel Feldman & Iván Dario Quiñones
建筑设计：Plan Padrino, Presidential counseling for early childhood, Presidency of the Republic of Colombia
总顾问：Maria Cristina Trujillo de Murioz
出资人：Gabriel Cano, Andrés Ortega, Eugenio Ortiz, Sandra Pineda
建造商：Fundación Compartir
甲方：Instituto Colombiano de Bienestar Familiar(ICBF)
用地面积：4,805m²
总建筑面积：1,823m²
施工时间：2012.10
竣工时间：2013.9
摄影师：courtesy of the architect

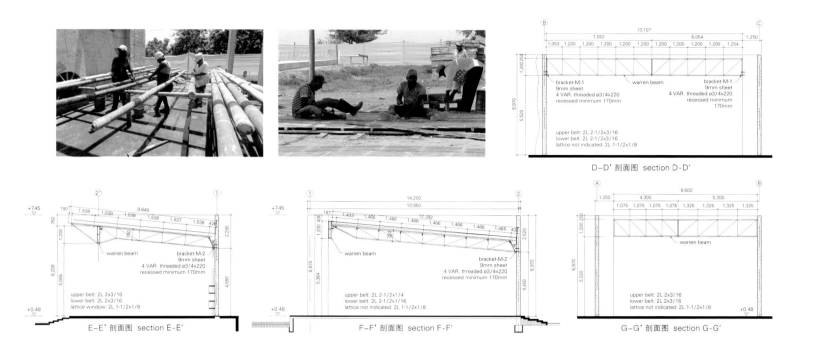

D-D' 剖面图 section D-D'

E-E' 剖面图 section E-E'

F-F' 剖面图 section F-F'

G-G' 剖面图 section G-G'

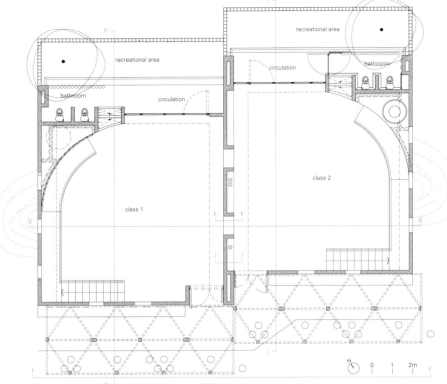

详图4 detail 4

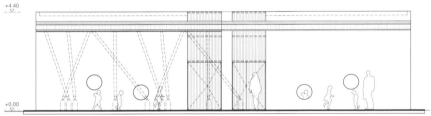

f-f' 剖面图 elevation f-f'

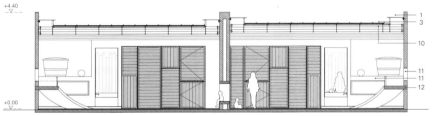

g-g' 剖面图 section g-g'

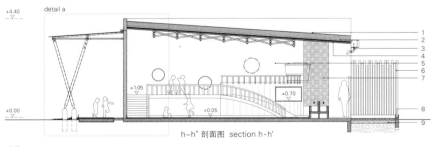

h-h' 剖面图 section h-h'

i-i' 剖面图 section i-i'

1. lateral polycarbonate skylights
2. metal mesh screen
3. roof 10% slope
4. metal sheet rain water gutter
5. bamboo ceiling
6. reinforced poured concrete wall
7. bamboo fence ø8cm
8. river pebbles 1cm
9. ceramic tile
10. bamboo structural beams
11. reinforced concrete slab
12. wall perforation ø65cm

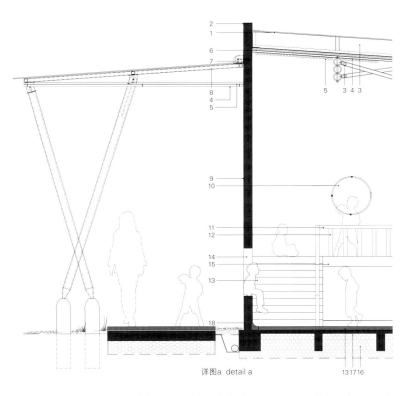
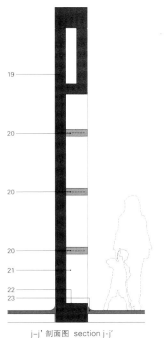
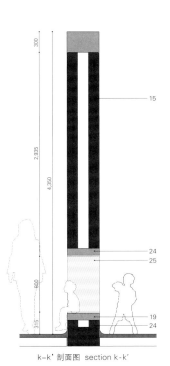

详图a detail a   j-j' 剖面图 section j-j'   k-k' 剖面图 section k-k'

1. lateral polycarbonate skylights
2. metal sheet protection
3. metal mesh screen
4. ceiling structure
5. bamboo ceiling
6. bamboo structural beams
7. joint-articulated
8. steel structure for roofing
9. reinforced poured concrete wall
10. wall perforation ø65cm
11. wood railing ø5cm
12. steel railing
13. poured concrete floor, polished and sealed
14. wall perforation ø65cm
15. reinforced concrete slab
16. compacted filling
17. concrete slab
18. curved baseboard in polished and sealed
19. reinforced poured concrete wall, horizontal split bamboo texture form
20. shelf 10cm reinforced concrete, grey
21. reinforced poured concrete wall, formaleta horizontal of bamboo mat
22. poured concrete supporting
23. curved baseboard in polished granite
24. tie beam
25. plaster inner side of the opening, variable colors

158

社区建筑——个性化建筑 In the Community – The Individuality and the Institution

木质牙科诊所
Kohki Hiranuma Architect & Associates

设计过程
design process

根据功能而设计的空间关系
relationship between spaces according to its function

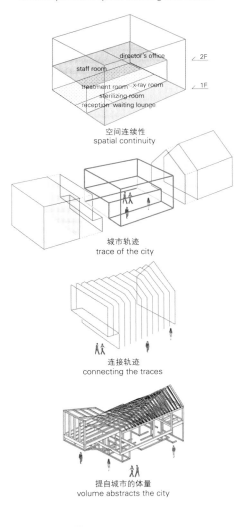

空间连续性
spatial continuity

城市轨迹
trace of the city

连接轨迹
connecting the traces

提自城市的体量
volume abstracts the city

作为城市背景和绿化区的立面
facade as the backdrop of the city and greenery

这个项目的场地以前是日本的土佐住宅博览会的所在地。这个住宅区保留了这处文化背景的宁静氛围,而主街两旁的樱花树也同样使人印象深刻。经过11个月对其位置和环境的研究,许多元素吸引了建筑师的注意,如醒目的空房子、枯树、便利的商店、药房、超市和其他周边区域内不相称的设施。因此,它逐渐被改造成一个独立的社区,强调了当今时代的顾客获得生活必需品的便利性。然而,这个医疗场所在融入现代小区的同时也保留住了这片安静住宅区留给人们的价值。建筑师打算通过开发社区来建造一座可持续的建筑,并且在展示环境的同时对环境进行重新美化。

这处场地是空置的,因此,从曾建于此的房子及其外围便可推测出体量的形成。该建筑最初仅始于模仿其西侧相邻住宅的三角形屋顶轮廓,之后轮廓转而延伸到东面,仿佛在表达现代主义运动,因一道曲线刻于其中。这座建筑看上去是现代的,但是在社区仍属罕见。这一构成体量的抽象方案仿佛使人们忘记了它的存在。这个城市的生命力由这座建筑反映出来,白色的帆布与樱花街交相辉映,而立面上的天然材料与周围环境水乳交融。

除了保留北面马路旁边的一株新植物外,其他植物皆置于人行道旁,那里同样能引人注目,因为玻璃提高了人们视野的清晰度,从而看到支撑整座建筑的木结构。建筑师拟定的计划是樱花街将内外空间混合成一个社区,而不仅仅是获得绿化率。"树木给人们带来勃勃生机和焕然一新的环境。"虽然只是一座小小的建筑,但是这种设计在留存其历史背景的同时也对城市文化起到了启发作用。

Timber Dentistry

The location of the project was formerly "the area" for Taisho Housing Expo. This residential area preserves the serene ambiance inherited from the cultural context that also has great impression of Sakura trees along the main street. After 11 months of research and studying the location and its environment, many things took the attention, such as the conspicuous empty houses, withered trees, convenient stores, pharmacy, supermarkets and other unsuitable facilities that surround the area. Hence, it gradually changes into a dependent community that emphasizes the convenience for the daily essentials to modern consumers. However, the presence of this medical facility conserves the remaining value in this quiet residential area, while blending into its modern community. The architects intended to realise a sustainable building liable in developing the community and represent its historical context while achieving environmental renovation.

西立面 west elevation

南立面 south elevation

东立面 east elevation

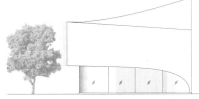

北立面 north elevation

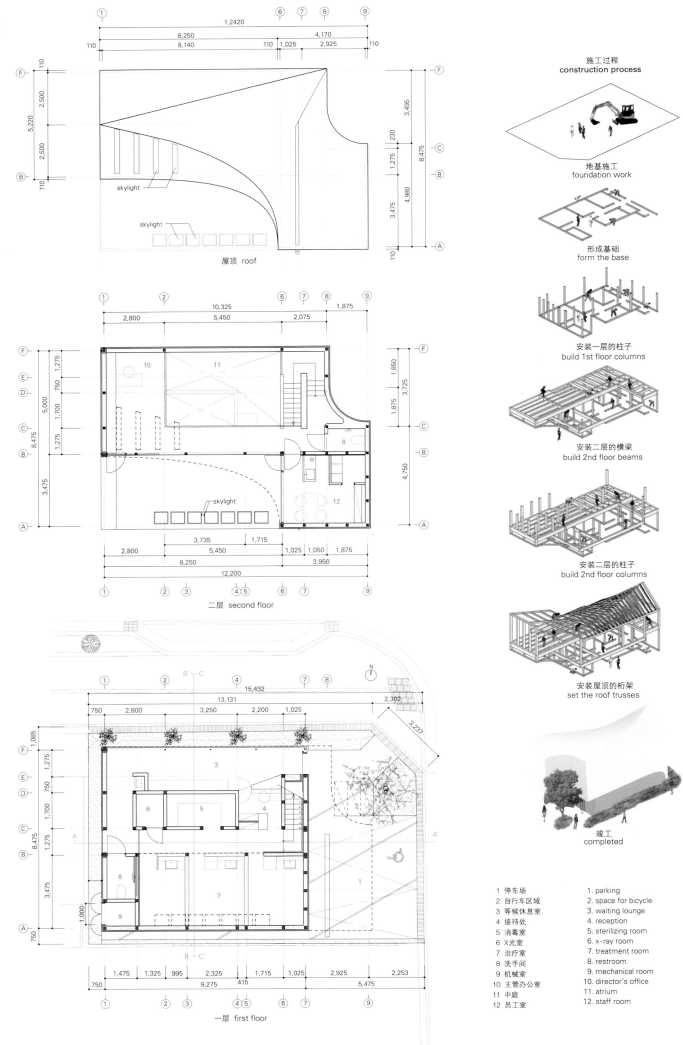

The site was vacant upon acquirement. Thus, the houses formerly standing on the site and its peripherals had been the clue to form the volume. It simply started on tracing the silhouette of the gabled-roof of its neighboring house on the west. The silhouette transforms towards the east, as if it is expressing the motion of modernism having a curve line in it. The building seemed to be modern yet rare to the community. This abstract solution of forming its volume obliterates its existence. The life of the city reflects on this architecture, where white canvass complements the street of Sakura whereas the natural materials on its facade blend into the environment.

Instead of conserving a new plant next to the road on the north, setting the plants right next to the sidewalk also captivates the attention as the glass increases clearness of the view to the timber structure that supports the whole architecture. The architects aimed a plan that street of Sakura blends the interior and exterior spaces into one community not just merely to achieve the green space ratio. "Trees, give brightness and cater for refreshing environment." In spite of being a small building, this approach stimulates the culture of the city while preserving its historical context.

项目名称：Timber Dentistry
地点：2-4-20 Sakuragaoka Mino City, Osaka
建筑师：Kohki Hiranuma Architect & Associates
设计团队：
Kohki Hiranuma _ Kohki Hiranuma Architect & Associates,
Masahiro Inayama _ Holzstr,
Shigeo Kawazoe _ Kawazoe Environmental Design,
Kohichi Minami _ Minami Building Equipment Design
合作设计：Reiko Sudo (NUNO)
施工工程师：Nishimura Architectural Studio
甲方：Teramura Dental Clinic
用途：Dental Clinic
用地面积：155.92m² / 总建筑面积：97.29m²
有效楼层面积：144.23m²
总建筑规模：two story
结构：wood
设计时间：2012.6—2013.9 / 施工时间：2013.10—2014.6
摄影师：©Satoshi Shigeta (courtesy of the architect)

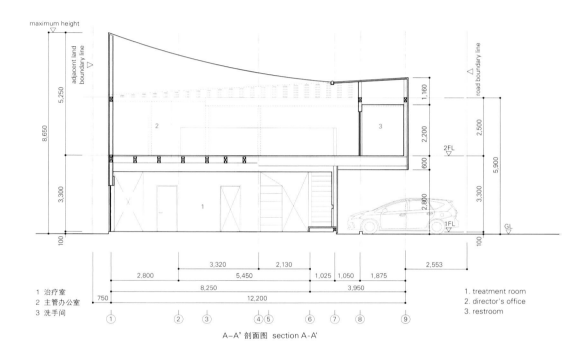

1 治疗室 / 1. treatment room
2 主管办公室 / 2. director's office
3 洗手间 / 3. restroom

A-A' 剖面图 section A-A'

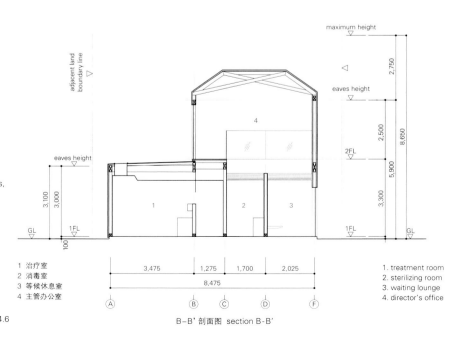

1 治疗室 / 1. treatment room
2 消毒室 / 2. sterilizing room
3 等候休息室 / 3. waiting lounge
4 主管办公室 / 4. director's office

B-B' 剖面图 section B-B'

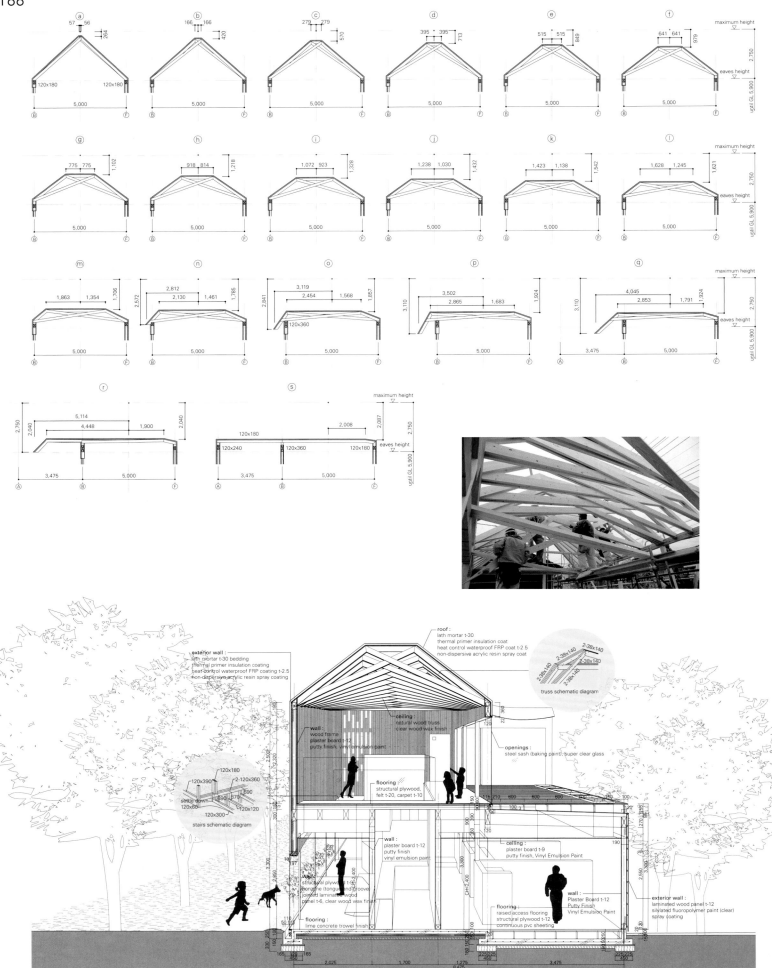

C-C' 剖面图 section C-C'

社区建筑——个性化建筑 In the Community – The Individuality and the Institution

哥本哈根癌症防治与康复中心
Nord Architects

患上癌症就像开始一场旅行,你并不知道哪里会是尽头。它要求人们拥有与疾病较量和接受癌症患者这一新身份的力量。研究表明建筑能对人体康复产生积极作用。人性尺度的设计和宜人的环境有助于患者康复。尽管这样,大多数医院并不让人舒适。哪怕是从接待处到餐厅的路都很难找。如果想让人们在我们的医院感觉更好,我们需要消除所谓的制度,创造更令人愉快的医疗环境。由哥本哈根的诺德建筑师事务所设计的癌症防治与康复中心便做到了这一点。

这幢建筑坐落在哥本哈根市中心附近,与哥本哈根大学医院(丹麦国立医院)在同一区域。这样患者在医院接受治疗后可以直接去康复中心。在马路的另一边是帕氏医学研究所。

哥本哈根的癌症防治与康复中心被看做是一座标志性建筑。它提醒人们增强抗癌意识,但又不对患者心理造成伤害。该中心由很多小房子组合到一起,提供了现代理疗环境所需的空间,同时又不降低个人舒适度。这些房子由高高的屋顶连接到一起,屋顶的形状就像日本的折纸艺术品。这种结构使该建筑成为独具特色的标志。

走进这幢建筑,你会发现自己置身于志愿者服务的舒适的休息大厅。从这里你可以前往房子的其他区域,包括一个用来静心冥想的庭院、几处锻炼的地方、一个可以学习烹饪健康食物的大众厨房以及患者会面的房间等。

Copenhagen Center for Cancer and Health

Getting cancer is like embarking on a journey, you don't know where it will end. It requires strength to cope with the disease and take on the new identity as a cancer patient. Research shows that architecture can have a positive effect on people's recovery from disease. A human scale and welcoming atmosphere can help people to get better. Despite of this, most hospitals are hardly comfy. Just finding the way from the reception to the canteen can be difficult. If we want people to get better at our hospitals, we need to deinstitutionalize and create a welcoming healthcare. The Center for Cancer and Health designed by Nord Architects Copenhagen does just that.

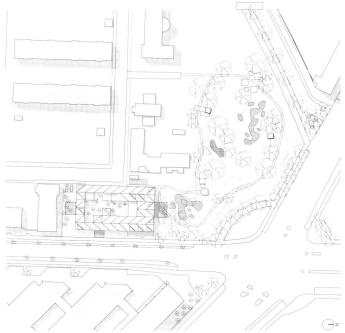

The building is situated close to the city center of Copenhagen in the same area as Copenhagen University Hospital(Rigshospitalet), so that patients can go to the healthcare center after their treatment at the hospital. On the other side of the road is the Panum Institute of Medicine.

The Center for Cancer and Health in Copenhagen is conceived as an iconic building, which creates awareness of cancer without stigmatizing the patients. Designed as a number of small houses combined into one, the center provides the space needed for a modern health facility, without losing the comforting scale of the individual. The houses are connected by a raised roof shaped like a Japanese paper art origami, which gives the building a characteristic signature.

Entering the building you find yourself in a comfy lounge area manned by volunteers. From here you move onto the others parts of the house, which include a courtyard for contemplation, spaces for exercises, a common kitchen where you can learn to cook healthy food, meeting rooms for patients groups etc..

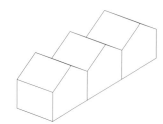

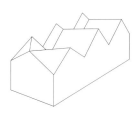

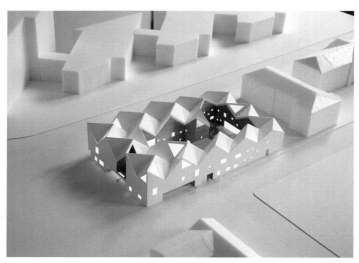

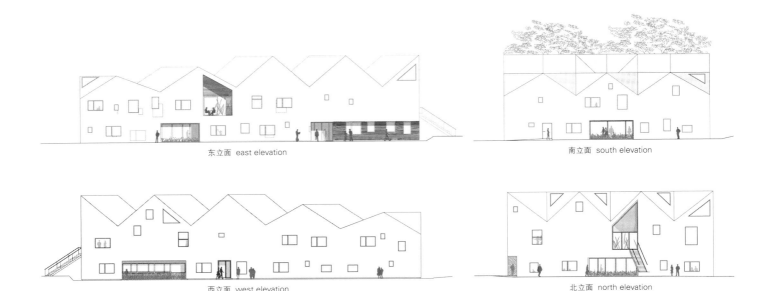

东立面 east elevation

南立面 south elevation

西立面 west elevation

北立面 north elevation

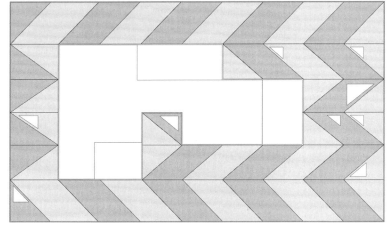
屋顶 roof

一层 first floor

二层 second floor

地下一层 first floor below ground

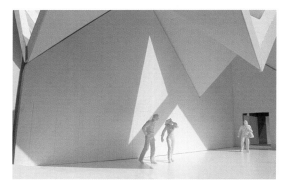

1 主入口	16 复印室	1. main entrance	16. copy room
2 办公室	17 小组房间	2. office	17. group room
3 休息室	18 仓库	3. lounge	18. depot
4 图书馆	19 行政区	4. library	19. administration
5 会客室	20 花园	5. meeting room	20. garden
6 信息墙	21 治疗室	6. info wall	21. treatment
7 厨房	22 秘书办公室	7. kitchen	22. secretary office
8 员工食堂	23 主管办公室	8. staff canteen	23. chief office
9 儿童角	24 志愿者办公室	9. children corner	24. office for volunteers
10 次入口	25 和平花园	10. secondary entrance	25. peaceful garden
11 信息台	26 露台	11. information	26. terrace
12 室外咖啡厅	27 体育场	12. outdoor cafe	27. gym
13 清洁室	28 更衣室	13. cleaning	28. changing room
14 衣柜	29 淋浴间	14. wardrobe	29. shower
15 卫生间		15. toilet	

A-A' 剖面图 section A-A'

B-B' 剖面图 section B-B'

1	办公室	1.	office
2	会客室	2.	meeting room
3	行政区	3.	adminstration
4	秘书办公室	4.	secretary office
5	体育馆	5.	gym
6	仓库	6.	depot
7	治疗室	7.	treatment
8	休息室	8.	lounge
9	图书馆	9.	library
10	露台	10.	terrace
11	和平花园	11.	peaceful garden
12	小组房间	12.	group room
13	淋浴间	13.	shower
14	更衣室	14.	changing room
15	主入口	15.	main entrance
16	室外咖啡厅	16.	outdoor cafe
17	复印室	17.	copy room
18	清洁室	18.	cleaning
19	厨房	19.	kitchen

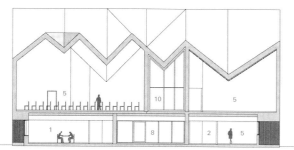
C-C' 剖面图 section C-C'

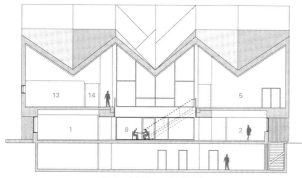
D-D' 剖面图 section D-D'

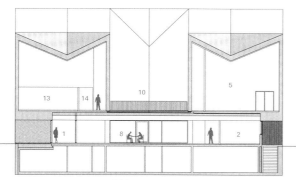
E-E' 剖面图 section E-E'

F-F' 剖面图 section F-F'

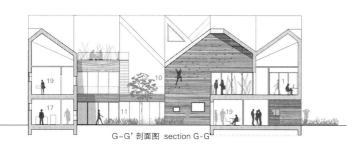
G-G' 剖面图 section G-G'

H-H' 剖面图 section H-H'

项目名称：Center for Cancer and Health
地点：Copenhagen, Denmark
建筑师：Nord Architects
工程师：Wessberg
景观建筑师：Nord Architects
甲方：Municipality of Copenhagen
甲方顾问：Moe & Brødsgaard
室外区域面积：439m²
总建筑面积：991m²
有效楼层面积：2,230m²
造价：DKR 56,000,000
施工时间：2009—2011
摄影师：©Adam Mørk (courtesy of the architect)

利迈健康中心

Atelier Zündel Cristea

位于利迈的健康中心是为精神、知觉和身体方面面临挑战的人们及其访客而设计的，是一座充满积极向上的氛围且安静的建筑。它将打破传统机构的形象，与家庭住宅的氛围尽量相似。

规划的空间通常与特殊的功能相关联，而这里的空间与其相反，以自然为主，在规划时汲取了轻松的氛围和一定的自由性。设计理念主要考虑了三点：首先是将建筑作为一种治疗工具，考虑到最终的用户，建筑空间必须与真实的世界不同。其次是受到庇护的花园是一处自由的空间：花园在门厅中发挥了重要的作用，因此项目在实和虚之间，换句话说，在室内外之间获得了平衡点（和不平衡点）。最后一点是将流线空间作为起居空间：精准的路线网络遍布不同区域，且不仅仅作为连接的功能。走廊根据其沿线的空间而进行延伸、转换和变化。材质面板和颜色也都将这些变化囊括其中。

在功能布局方面，建筑有明确且人性化的空间布局，且带有可变化的空间层次，在未来具有一定的灵活性。项目包含一座设有行政和服务区的建筑以及四座用于住宿的翼楼，翼楼与其他的起居空间（起居室、休闲室、护理室以及餐厅等等）相连接。该平面为带有两个主轴的长方形框架。南北向轴线主要体现为第一座建筑内的主要员工流线，而东西轴线则为住户和物流提供通道。每个用于住宿的翼楼都包含14间单独的房间，且围绕着一个室内院子而建，院子提供自然照明。除了上述提及的

四个院子之外，还有两座大型公共花园，面积均为350m²，另外还有一座大型开放式花园，所有人都可进入，位于场地的东部。

就技术方面而言，该项目兼顾了生物气候和环境方面的原则。建筑为混凝土结构，以获得绝佳的热惯性和隔音效果。绿色屋顶致力于提高热惯性性能，且在晴天，太阳能保温板能够供应60%的热水使用。屋顶的一部分在北侧升起，使起居空间产生漫射光和自然通风。最后，传统的自然材料（木材和油毡）用来覆盖室内外空间。室外的木墙板覆在加强绝缘层上，完全消除了热桥现象，而某些室内空间使用的木材也能够形成温馨且舒缓的氛围。

Medical Care Center in Limay

The Medical Care Center in Limay, especially designed for mentally, sensorially and physically challenged people and their visitors, is an optimistic and serene building, which breaks with the traditional image of the host institution to become as close as possible to that of a family home.

Contrary to programmed spaces, which are associated with specific functions, spacing pervaded by nature acquires a relaxing breath and a little freedom in defining its possible spaces. The design concept considers three main points. The first point is the building as a therapeutic tool: considering the end users, the building spaces have to be as heterogeneous as the real world. The second point is the protected garden as a space of freedom: the garden plays an essential role in the Foyer, and for that the project engages in an equilibrium (and disequilibrium) between the full and the empty, in other words between interior and exterior spaces. The last point is the circulation space as living space: the precise network that irrigates the different spaces is not merely reduced to the function of a link. The passageway expands, transforms and varies according to the space it runs alongside. The palette of materials and colors also contain this diversity.

With regard to the organization of the functions, the building has a clear and affective configuration of the spaces, with a spatial flux hierarchy and a certain level of future flexibility. The architectural project consists in an administrative and service block and four

1 花园 2 大厅 1. garden 2. hall
A-A' 剖面图 section A-A'

1 休闲室 2 起居室 1. relaxation room 2. living room
B-B' 剖面图 section B-B'

1 花园 2 餐厅 3 护理室 4 活动室 1. garden 2. dining room 3. care room 4. sport room
C-C' 剖面图 section C-C'

1 大厅 2 花园 3 卫生和餐饮服务区 4 卧室 1. hall 2. garden 3. hygiene and food service 4. bedroom
D-D' 剖面图 section D-D'

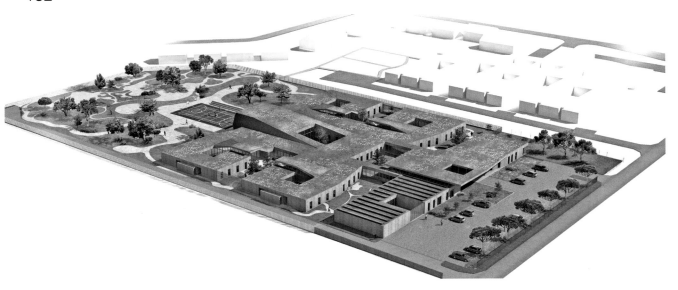

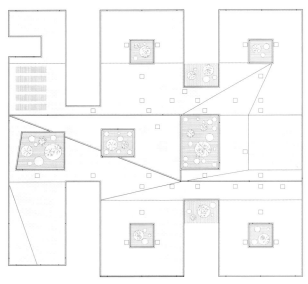

屋顶 roof

项目名称：Foyer d'Accueil Médicalisé in Limay
地点：Limay, France
建筑师：Atelier Zundel Cristea
项目团队：Irina Cristea, Gregoire Zundel
合作者：Batiserf Ingenierie, Louis Choulet,
Bureau Michel Forgue, Infra Services
顾问：Bureau Michel Forgue, BET Louis Choulet, Batiserf Ingénierie
甲方：SIEHVS + SARRY 78
用地面积：20,000m²
总建筑面积：2,555m²
造价：EUR 7.5m
设计时间：2008.9
施工时间：2010.1
竣工时间：2011.10
摄影师：©Sergio Grazia(courtesy of the architect)

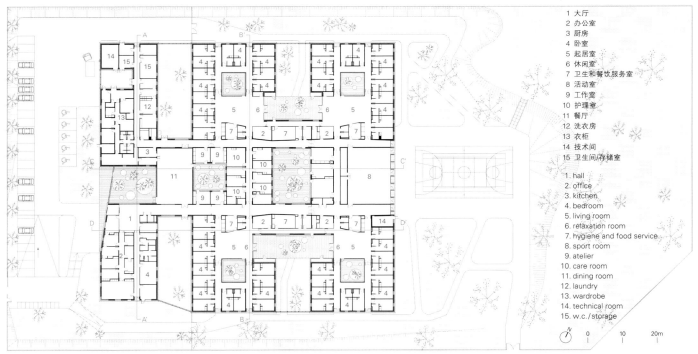

一层 first floor

1 大厅
2 办公室
3 厨房
4 卧室
5 起居室
6 休闲室
7 卫生和餐饮服务室
8 活动室
9 工作室
10 护理室
11 餐厅
12 洗衣房
13 衣柜
14 技术间
15 卫生间/存储室

1. hall
2. office
3. kitchen
4. bedroom
5. living room
6. relaxation room
7. hygiene and food service
8. sport room
9. atelier
10. care room
11. dining room
12. laundry
13. wardrobe
14. technical room
15. w.c./storage

accommodation wings connected to other living services (living rooms, relaxation rooms, care rooms, dining room, etc.). The plan follows a rectangular framework with two main axis. The North-South axis is dedicated to the circulation of the staff in the first block; the East-West double axis permits the circulation of residents and the logistics. Each accommodation wing, composed of 14 single rooms, is developed around an internal court with the function of natural lighting. In addition to the above-mentioned four courts, two large communal gardens of 350m² are present. A large open garden, accessible to all, is located in the eastern part of the grounds.

As far as technical aspects concerned, the project considers several bioclimatic and environmental principles. The building structure is in concrete, in order to obtain excellent thermal inertia and acoustic insulation. The green roof contributes to improve the thermal inertia and the thermal solar panels are able to provide the 60% of hot water during sunny days. Part of the roof, raised toward the north side, gives a diffused light and natural ventilation to living rooms. Finally, traditional natural materials (wood and linoleum) are used for covering external and internal spaces. The exterior wood siding, positioned on a reinforced insulation, eliminates completely the thermal bridges and the use of wood in some internal spaces contributes to make a soft and soothing atmosphere.

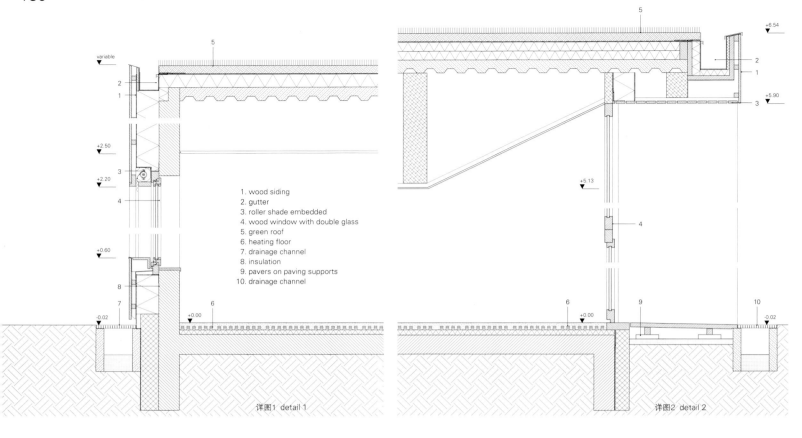

1. wood siding
2. gutter
3. roller shade embedded
4. wood window with double glass
5. green roof
6. heating floor
7. drainage channel
8. insulation
9. pavers on paving supports
10. drainage channel

详图1 detail 1   详图2 detail 2

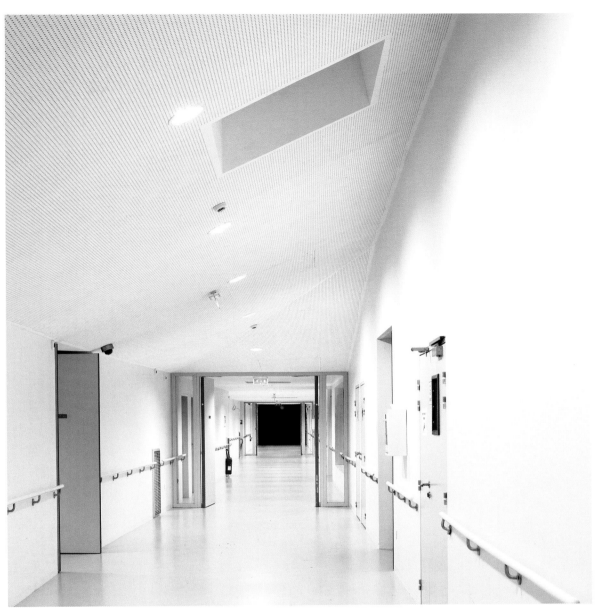

>>106

Dorte Mandrup Arkitekter
Is an international practice, based in Copenhagen, Denmark, founded by Dorte Mandrup in 1999. Is renowned for original architectural design of high standards, both conceptually and formally from the scheme to the 1:1 detailing. Pushes the boundaries for known typologies, working in the wide spectrum of architecture but with the specific aim to solve programs in new and aesthetic ways. This had led to appraised buildings, often used as references for visionary solutions. Is responsible towards the client for projects in their entirety from schematic design, detailing, cost management to site supervision. Is used to lead big teams for complex building projects, focusing on involving everyone in the team for a constructive dialogue.

>>22

Andrew Berman Architect
Andrew D. Berman majored studio art, architecture and art history at Yale College. After graduation in 1984, he studied at the School of Architecture, Yale University and received M.Arch in 1988. His projects have received numerous state and local AIA design excellence awards including Architecture Merit Award, Merit in Design Award and Design Excellence Award. In 2009, The Architectural League of New York named him an "Emerging Voice". Is registered architect of New York State, New Jersey, Maine and member of various organizations including AIA(American Institute of Architects), Architectural League of New York and The Storefront for Art and Architecture.

>>114

Vo Trong Nghia Architects

Vo Trong Nghia graduated from Nagoya Institute of Technology with a B.Arch in 2002 and received Master of Civil Engineering from Tokyo University in 2004. In 2006, he established Vo Trong Nghia Co.Ltd.

>>54

Estudio Arquitectura Hago

A spanish office based in Madrid, established in 2005 by Emilio Delgado-Martos and Antonio Álvarez-Cienfuegos Rubio. After nearly 10 years of work and more than 90 projects, they are specialists in restoration, cultural spaces and heritage. Has designed a wide range of projects from the small scale of a private home to the large scale of urban design. Has been nominated for the European Union Prize for Contemporary Architecture-Mies van der Rohe Award 2015. Antonio Álvarez-Cienfuegos Rubio has worked in several offices such as Cruz y Ortiz Arquitectos, Alberto Campo Baeza and Nuñez Ribot Arquitectos Asociados. Emilio Delgado Martos is teaching at Graduate School of Architecture and Faculty of Fine Arts & Design in Francisco de Vitoria University (UFV), Madrid since 2006.

>>178

Atelier Zündel Cristea

Was founded in 2001 by Irina Cristea[left] and Grégoire Zündel[right] who graduated from the National Superior school of Arts(ENSA) in Strasbourg, France in 1995. Irina Cristea was born in Bucharest, Romania. Has collaborated with M. Fuksas, Du Besset-Lyon and Hsin Yieh after graduation. Grégoire Zündel was born in Colmar, France. Has collaborated with M. Fuksas, J. Ferrier and Terry Farrell & Partners after graduation. Has taught at his alma mater as visiting teacher for two years.

>>12

MGS Architects

Eli Giannini received B.Arch and M.Arch from Royal Melbourne Institute of Technology(RMIT). Has been working at McGauran Giannini Soon Pty Ltd. Since 1989. Has contributed to the architecture profession at Royal Australian Institute of Architects, Victorian Chapter Awards Task Force and the 2007 National Conference organizing committee. Was made a Lifetime Fellow of the AIA(LFAIA) in 2008. Has worked in RMIT as an examiner in the Masters program and the chair of the Course Advisory Committee of the Faculty of Design. Joshua Wheeler studied Architecture and Building Science at Victoria University of Wellington. Is a Registered Architect of AIA(Australian Institute of Architects), NZIA(New Zealand Institute of Architects), RIAI(Royal Institute of the Architects of Ireland) and California. Has been working at McGauran Giannini Soon Pty Ltd. Since 2009. Is a Director with experience in design, planning and architectural contract administration as well as urban design and master planning.

>>76

TAO

Was founded in 2009 by Hua Li who received a B. Arch. from Tsinghua University in 1994 and M. Arch. from Yale University in 1999. Is a Beijing based design studio committed to architecture, urban, landscape, and furniture design. With their most projects positioned in specific cultural and natural settings in China, TAO makes architecture deeply rooted in its social and environmental context with respect of its gravity. The issues such as the sense of place, response to climate, efficient use of local resource, appropriate material and construction method are always explored in every project responding to its specific context.

>>126

Mazzanti Arquitectos

Giancarlo Mazzanti was born in Barranquilla, Colombia in 1963. Graduated with a degree in Architecture from the Pontifical University of Javeriana in Bogotá, Colombia in 1987. Received a postgraduate degree in history and theory of architecture and industrial design from the University of Florence, Italy in 1991. Has taught in several Columbian Universities and some of the most prestigious American Universities such as Princeton University and Harvard Graduate School of Design (GSD). Has been the winner at the XX Colombian Architecture Biennial, Ibero American Biennial, Panamerican Architecture Biennial. Also won the Global Award for Sustainable Architecture prize from the French Institute of Architecture.

>>136

5+1 AA Alfonso Femia Gianluca Peluffo

Alfonso Femia[middle] and Gianluca Peluffo[left] founded 5+1 in 1995 and created 5+1AA in 2005. In 2006, Simonetta Cenci[right] became a partner of 5+1AA and they opened a studio in Milan. They deal with the theme of simultaneity in the relationship between city, territory and architecture, constructing it in the form of reality. In 2011, they won the Philip

©Sharyn Cairns

>>46
Brewster Hjorth Architects
Ian Brewster is a Director and CEO of BHA that he jointly founded in 1984. Received B.Arch from the University of New South Wales(UNSW) with University Medal. Has particular interests in the incorporation of ESD(environmental) Design principles in BHA projects and the incorporation of Digital Information systems in modern public buildings. Luigi Staiano is Senior Associate of BHA. Received B.Arch from the University of Sydney and joined BHA in 1994 after working on most of BHA's heritage and masterplan projects, as well as many civic, community and university projects. Maria Colella is Assistant Project Architect of BHA. Graduated from the University of Sydney and joined BHA in 2004. Has become Design team leader through her dedicated work on many projects.

>>66
Waterford City Council Architects
Rupert Maddock has received B.Arch in 1981 from the University College Dublin(UCD) and studied at the University of Pennsylvania(MLA, 1983). Bartosz Rojowski and Agnieska Rojowska received their diplomas in architecture from Cracow University of Technology in 2001. Ali Jay studied Architecture at Canterbury School of Architecture, Kent and received her B.A hons degree in 1995. Bartosz and Agnieska have now set their own practice ROJO-Studio Architects in Waterford, Ireland. Sculptor Stephen Burke studied fine art in Dublin and Cork and Architectural Stone Carving in London. They have received prestigious awards such as Irish Architecture Award and UK Stone Federation Award.

>>168
Nord Architects
Is a consultancy practice and member of DANSKE ARK/DI that work with socially engaged projects in the built environment primarily throughout Scandinavia. Management team consists of founding partners Morten Rask Gregersen[right] and Johannes Molander Pedersen[left], and office managers Tine W. Holmboe and Mia Baarup Tofte. They have specialized experience with sustainable urban development and technical building solutions within different typologies at various scales. On this process, they focus on energy-efficient and optimized solutions, resources, climate control, acoustics, environment and optimal operating materials. Have a unique knowledge in several areas such as strategic urban development plans, public space design, day care and school facilities, sports facilities and culture facilities. They believe that good architecture creates synergy for all involved - namely society, users and entrepreneurs.

>>158
Kohki Hiranuma Architect & Associates
Was established by Kohki Hiranuma. He was born in Osaka, Japan, 1971 and studied architecture at AA School in London. Awards inside the country follow; Japan Federation of Architects & Building Engineers Association Award, AIJ Award, JCD Semi-Grand Award, JID Award, Japan Architecture Renovation and Conversion Gold Prize, and Japan Industrial Association Good Design Award. In Foreign countries, he received Grand Design International Architecture Award (England), Innovative Architecture International Building Award (Italy), and German Design Award (Germany). His work has been exhibited in the National Museum of Art in Osaka, Japan in 2009 and Venice Biennale International Architecture Exhibition, Italy in 2014. Currently he teaches at Osaka University and leads Non-Profit Orgarnization/Art & Architect Festa(NPO/AAF).

Douglas Murphy
Studied architecture at the Glasgow School of Art and the Royal College of Art, completing his studies in 2008. As a critic and historian, he is the author of The Architecture of Failure(Zero Books, 2009), on the legacy of 19th century iron and glass architecture, and the forthcoming Last Futures (Verso, 2015), on dreams of technology and nature in the 1960s and 1970s. Is also an architecture correspondent for Icon Magazine, and writes regularly for a wide range of publications on architecture and culture.

Alison Killing
Is an architect and urban designer based in Rotterdam, the Netherlands. Has written for several architecture and design magazines in the UK, contributing features and reviews to Blueprint and Icon and editing the research section of Middle East Art Design and Architecture. Most recently, she has worked as a correspondent for the online sustainability magazine Worldchanging. Has an eclectic design background, ranging from complex geometry and structural engineering, to humanitarian practice, to architecture and urban design and has worked internationally in the UK and the Netherlands, but also more widely in Europe, Switzerland, China and Russia.

Arkitektfirma Helen & Hard AS
Was founded in Stavanger, Norway by Siv Helene Stangeland and Reinhard Alois Kropf in 1996. They are working with 20 staff from 8 different countries including their architects, Håkon Solheim and Randi Augenstein with offices in both Stavanger and Oslo. Siv Helene Stangeland was born in Stavanger, Norway in 1966. Studied at the Oslo School of Architecture and Design(AHO) and the Barcelona School of Architecture(ETSAB). Has taught at the Norwegian University of Science and Technology(NTNU), Chalmers University of Technology, Royal Institute of Technology(KTH) and AHO. Reinhard Alois Kropf was born in Gleisdorf, Austria in 1967. Studied at the Graz University of Technology(TU Graz), Austria and AHO. Has taught at the Superior School of Architecture in Paris(ESA) and AHO. Håkon Solheim has studied at Trondheim and Helsinki University of Technology(NTNU). Randi Augenstein has studied at Berlin University of the Arts and the Royal Danish Academy of Fine art, School of Architecture.

>>90

Cox Rayner Architects
Michael Rayner is Principal Director of Cox Rayner Architects and an Executive Director of Cox Architecture Pty Ltd. Since establishing Cox Rayner in Queensland in 1990, he has led the practice to international standing. Is currently teaching at University of Queensland and Griffith University. As a senior Project Leader, Justin Bennett has joined Cox Rayner Architects in 1992. Has been responsible for the project delivery of projects from design development to completion. Has strengths in the design, contract documentation and site administration of large complex projects. Casey Vallance graduated from the University of Queensland with B.Arch in 2002 and joined Cox Rayner Architects in January 2003. In 2007, he became an Associate, and in 2010 he became the youngest ever director of this group. Received numerous awards such as World Architecture Festival Award and AIA Public Architecture Award.

>>144

Daniel Feldman & Iván Dario Quiñones
Daniel Feldman[right] and Iván Dario Quiñones[left] worked as lead architects for the High Presidential Advisor for Early Childhood where they were in charge of designing and delivering the community outreach strategy, participatory workshops, project design, construction oversight, and financing strategy for the projects to be built. By working as architects within the government they were able to re-imagine the processes of construction of public infrastructure and developed a framework for the execution of design and construction of high quality public buildings in secluded and war torn settings. They worked in tandem with the Advisors staff in order to build in the economic, sociological, anthropological, and legal structures necessary for the projects to be built.

C3, Issue 2015.1
All Rights Reserved. Authorized translation from the Korean-English language edition published by C3 Publishing Co., Seoul.

© 2015 大连理工大学出版社
著作权合同登记06-2015年第09号
版权所有·侵权必究

图书在版编目(CIP)数据

社区建筑：汉英对照 / 韩国C3出版公社编；
党振发等译著. —大连：大连理工大学出版社，2015.3
（C3建筑立场系列丛书）
书名原文：C3 In the Community
ISBN 978-7-5611-9793-6

Ⅰ. ①社… Ⅱ. ①韩… ②党… Ⅲ. ①社区－建筑设计－汉、英 Ⅳ. ①TU984.12

中国版本图书馆CIP数据核字(2015)第055782号

出版发行：大连理工大学出版社
　　　　　（地址：大连市软件园路80号　邮编：116023）
印　　刷：上海锦良印刷厂
幅面尺寸：225mm×300mm
印　　张：12
出版时间：2015年3月第1版
印刷时间：2015年3月第1次印刷
出 版 人：金英伟
统　　筹：房　磊
责任编辑：许建宁
封面设计：王志峰
责任校对：高　文

书　　号：978-7-5611-9793-6
定　　价：228.00元

发　　行：0411-84708842
传　　真：0411-84701466
E-mail：12282980@qq.com
URL：http://www.dutp.cn